Microwave-Assisted Polymerization

RSC Green Chemistry

Editor-in-Chief:
Professor James Clark, *Department of Chemistry, University of York, UK*

Series Editors:
Professor George A. Kraus, *Department of Chemistry, Iowa State University, Ames, Iowa, USA*
Professor Andrzej Stankiewicz, *Delft University of Technology, The Netherlands*
Professor Peter Siedl, *Federal University of Rio de Janeiro, Brazil*

Titles in the Series:
 1: The Future of Glycerol: New Uses of a Versatile Raw Material
 2: Alternative Solvents for Green Chemistry
 3: Eco-Friendly Synthesis of Fine Chemicals
 4: Sustainable Solutions for Modern Economies
 5: Chemical Reactions and Processes under Flow Conditions
 6: Radical Reactions in Aqueous Media
 7: Aqueous Microwave Chemistry
 8: The Future of Glycerol: 2nd Edition
 9: Transportation Biofuels: Novel Pathways for the Production of Ethanol, Biogas and Biodiesel
10: Alternatives to Conventional Food Processing
11: Green Trends in Insect Control
12: A Handbook of Applied Biopolymer Technology: Synthesis, Degradation and Applications
13: Challenges in Green Analytical Chemistry
14: Advanced Oil Crop Biorefineries
15: Enantioselective Homogeneous Supported Catalysis
16: Natural Polymers Volume 1: Composites
17: Natural Polymers Volume 2: Nanocomposites
18: Integrated Forest Biorefineries
19: Sustainable Preparation of Metal Nanoparticles: Methods and Applications
20: Alternative Solvents for Green Chemistry: 2nd Edition
21: Natural Product Extraction: Principles and Applications
22: Element Recovery and Sustainability
23: Green Materials for Sustainable Water Remediation and Treatment
24: The Economic Utilisation of Food Co-Products
25: Biomass for Sustainable Applications: Pollution Remediation and Energy
26: From C–H to C–C Bonds: Cross-Dehydrogenative-Coupling
27: Renewable Resources for Biorefineries
28: Transition Metal Catalysis in Aerobic Alcohol Oxidation
29: Green Materials from Plant Oils

How to obtain future titles on publication:
A standing order plan is available for this series. A standing order will bring
delivery of each new volume immediately on publication.

For further information please contact:
Book Sales Department, Royal Society of Chemistry, Thomas Graham House,
Science Park, Milton Road, Cambridge, CB4 0WF, UK
Telephone: +44 (0)1223 420066, Fax: +44 (0)1223 420247
Email: booksales@rsc.org
Visit our website at www.rsc.org/books

Microwave-Assisted Polymerization

Anuradha Mishra
Gautam Buddha University, Gautam Budh Nagar, India
Email: anuradha_mishra@rediffmail.com

Tanvi Vats
Gautam Buddha University, Gautam Budh Nagar, India
Email: tanvi.vats@gmail.com

James H. Clark
University of York, York, UK
Email: james.clark@york.ac.uk

RSC Green Chemistry No. 35

Print ISBN: 978-1-78262-317-5
PDF eISBN: 978-1-78262-318-2
ISSN: 1757-7039

A catalogue record for this book is available from the British Library

Published by The Royal Society of Chemistry,
Thomas Graham House, Science Park, Milton Road,
Cambridge CB4 0WF, UK

Registered Charity Number 207890

For further information see our web site at www.rsc.org

Printed in the United Kingdom by CPI Group (UK) Ltd, Croydon, CR0 4YY, UK

Preface

One of the greatest challenges facing humankind is the increasingly rapid depletion of natural resources. Whether it is oil from the sea or the land, or minerals from the ground, our insatiable appetite for consumption of resources and the inexorable growth in the world population puts us on a road to disaster – unless we adopt a more sustainable approach to resource consumption. Chemistry – which dictates the ways we can make all the articles we want in modern society, starting from petroleum and crude minerals as feedstocks, is at the heart of the modern industrial society and must be at the heart of the technological revolution we surely need. Critical to this revolution will be changes to the ways we do chemical processes. We need to carry out our chemistry with a more efficient use of resources and a reduction in the amount of waste. One vastly important step change will be the introduction of low energy technologies. We continue to use conventional heating techniques in chemistry but the chemical industry is a very large consumer of energy and must follow the trends in other sectors and seek to employ more efficient and low-carbon heating techniques. Alternative techniques can also help us to make chemical processes more efficient and generate fewer by-products, for example, by avoiding uneven heating that can cause hot spots and hence partial decompositions or side-reactions.

Since their development during the mid-20th century, synthetic polymers have become extensively used in almost every application imaginable. Some applications, such as medical, clothing, electrical, transport, construction and packaging, have become increasingly dependent on the lightweight, durable, reproducible and low cost characteristics of synthetic polymers. By the beginning of this decade, the production of synthetic polymers was 280 million tonnes, which accounts for 80% of the total production of the chemical industry. The identity, feedstock and composition of future polymers is under major review, especially as we move towards more bio-based

RSC Green Chemistry No. 35
Microwave-Assisted Polymerization
By Anuradha Mishra, Tanvi Vats and James H. Clark
© Anuradha Mishra, Tanvi Vats and James H. Clark 2016
Published by the Royal Society of Chemistry, www.rsc.org

products, but for any polymer the route for production and processing should be analyzed and optimized to ensure genuine energy efficiency, minimum waste and overall sustainability.

Microwave technology has gained acceptance as a mild and controllable energy tool, allowing simple and rapid processing. Industrial-scale treatment of food and materials at temperatures below 200 °C has been established as continuous tonne-scale processes with increased process selectivity. Microwaves are also widely used as laboratory processors, where countless researchers have discovered the beneficial effects of microwaves in helping to make chemical processes cleaner, more controlled and, in particular, quicker. Numerous publications have reported how, by switching from conventional to microwave heating, reactions have been completed in minutes rather than hours. The energy efficiency of microwave processes is more controversial, but it has been shown through careful energy measurements that microwave-assisted chemical reactions can be more energy efficient, especially for reactions that normally take many hours. While most microwave chemistry has stayed at the laboratory bench scale, some has made it to a larger scale, for example, in some chemical manufacturing processes and for the treatment of biomass both to enhance its energy density and its use as a source of organic chemicals. Given the enormous importance of polymer manufacturing, it seems very important that we seek to introduce energy- and time-efficient microwave processing to the industry.

This book seeks to fill the gap between the now quite widespread use of microwave processing in small molecule chemistry and biomass processing, and that of polymer synthesis and manufacture. The book starts with a consideration of the important points about carrying out microwave processing including the different types of reactors available and the particular hazards the operator needs to be aware of. The rest of the book is dedicated to the application of microwaves to polymer synthesis and manufacture. Major types of polymer syntheses including free-radical processes and ring-opening polymerizations are considered in detail. A wide range of polymers is considered, including widely used poly(ether)s and poly(amide)s. Separate chapters on synthesis of conducting polymers and formation of hydrogels are also included. The book goes further. An important aspect of polymer chemistry is how to modify polymers, for example, through incorporation into more complex structures like composites. Microwaves can, for example, be used to promote crosslinking and accelerate curing as well as derivatization of natural polymers. Finally, one chapter is dedicated to the widely studied area of peptide synthesis.

The enormous and growing scale of the polymer industry and the vast resource consumption it represents means that it must take precedence in undergoing the resource-efficiency make-over many manufacturing processes are being subject to. Microwaves can and should play an important role in this as we strive towards a more efficient and sustainable consumer society.

Anuradha Mishra
Tanvi Vats
James H Clark

Contents

RSC Green Chemistry No. 35
Microwave-Assisted Polymerization
By Anuradha Mishra, Tanvi Vats and James H. Clark
© Anuradha Mishra, Tanvi Vats and James H. Clark 2016
Published by the Royal Society of Chemistry, www.rsc.org

Microwave Radiations: Theory and Instrumentation

1.1 Introduction

Microwaves (MW) are electromagnetic waves whose frequencies range from 1 GHz to 1000 GHz. The higher frequency edge of microwaves borders on the infra-red and visible-light regions of the electromagnetic spectra. This explains why microwaves behave more like rays of light than ordinary radio waves do. It is because of this unique property that MW frequencies are classified separately from radio waves. As stated above, microwaves are electromagnetic waves, hence, in order to understand the properties of MW we need to have an understanding of the electromagnetic spectra.

Electromagnetic spectra can be defined as an arrangement of electromagnetic radiations in the order of their energy (which in turn is governed by their frequency or wavelength). Energy associated with each segment of the spectra is capable of producing a characteristic effect on the molecules exposed to them. Table 1.1 depicts the major regions of electromagnetic spectra and their effects.[1,2]

1.2 Microwave Effects

Microwaves are widely used for heating purposes. They have carved a niche as a non-conventional energy source in organic synthesis. Accelerated reactions, higher yields and milder reaction conditions make microwave assisted reactions stand apart. The supremacy of microwave irradiations can't be explained merely by rapid heating but by an overall "microwave effect" which encompasses thermal and non-thermal effects.[3,4] These effects are discussed in the next section of this chapter. The major points of difference

RSC Green Chemistry No. 35
Microwave-Assisted Polymerization
By Anuradha Mishra, Tanvi Vats and James H. Clark

Published by the Royal Society of Chemistry, www.rsc.org

Table 1.1   Major regions of electromagnetic spectra.

Region	Wavelength (Angstroms)	Frequency Hz	Energy eV	Effects
Radio	$>10^9$	$<3\times10^9$	$<10^{-5}$	Collective oscillation of charge carriers in bulk material (*plasma oscillation*)
Microwave	10^9–10^6	3×10^9–3×10^{12}	10^{-5}–0.01	Plasma oscillation, molecular rotation
Infra-red	10^6–7000	3×10^{12}–4.3×10^{14}	0.01–2	Molecular vibration, plasma oscillation (in metals only)
Visible	7000–4000	4.3×10^{14}–7.5×10^{14}	2–3	Molecular electron excitation (including pigment molecules found in the human retina), plasma oscillations (in metals only)
Ultra-violet	4000–10	7.5×10^{14}–3×10^{17}	3×10^3	Excitation of molecular and atomic valence electrons, including ejection of the electrons (*photoelectric effect*)
X-rays	10–0.1	3×10^{17}–3×10^{19}	10^3–10^5	Excitation and ejection of core atomic electrons, *Compton scattering* (for low atomic numbers)
Gamma rays	<0.1	$>3\times10^{19}$	$>10^5$	Creation of *particle–antiparticle pairs*. At very high energies a single photon can create a shower of high-energy particles and antiparticles upon interaction with matter.

between microwave heating and conventional heating are summarized in Table 1.2.

1.2.1   Thermal Effects

Thermal effects can be assumed to result from dipole–dipole interactions of polar molecules with electromagnetic radiations. They originate in the dissipation of energy into heat as an outcome of agitation and intermolecular friction of molecules when dipoles change their mutual orientation at each

Table 1.2   Characteristics of microwave and conventional heating.

Microwave heating	Conventional heating
Energetic coupling	Conduction/convection
Coupling at the molecular level	Superficial heating
Rapid	Slow
Volumetric	Superficial
Selective	Non selective
Dependent on the properties of the material	Less dependent

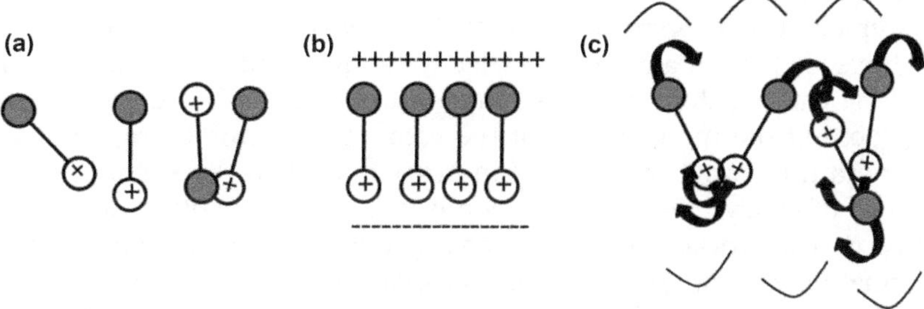

Figure 1.1   Polarization of molecules. (a) In the absence of external electric field. (b) In the presence of continuous electric field. (c) In the presence of high frequency alternating electric field.

alternation of electric field at very high frequency, as depicted in Figure 1.1. This energy dissipation in the core of materials allows a much more regular repartition in temperature when compared to classical heating.[5] The thermal effects manifest themselves in several forms, which are discussed in this segment.

For solids and semiconductors, charge space polarization is of prime importance concerning the presence of free conduction electrons, which are necessary for the microwave heating of solids.[6] In the case of liquids/solvents, only polar molecules absorb microwave radiations; non-polar molecules are inert to microwave dielectric loss. Effective microwave absorption results in higher boiling point values as compared to conventional heating. This phenomenon is called the "super heating effect". The super heating effect, also sometimes referred to as the overheating effect, can be explained by inverted heat transfer *i.e.* from an irradiated medium towards the exterior, as boiling nuclei are formed at the surface of the liquid. This effect explains the enhancement of reaction rates, higher efficiency and greater yields in organic and organometallic syntheses.[5,6]

Inhomogeneous heating or **thermal hotspots** have been detected in several microwave reactions. This is a thermal effect that arises due to inhomogeneity of the applied field, resulting in the temperature in certain zones within the sample being much greater than the macroscopic temperature. Hotspots may be created by the difference in dielectric properties of materials, by the uneven distribution of electromagnetic field strength or by volumetric dielectric heating under microwave conditions.[7]

The discussion of thermal effects of microwaves is incomplete without mentioning the **selective mode** of heating. MWs are a selective mode of heating in the sense that they exclusively interact with polar molecules. This characteristic has been exploited in solvents, catalysts and reagents. Selective heating has been used in two-phase solvent systems. Due to the differences in the dielectric properties of the solvents, different temperatures of the component phases can be attained. This effect can be of prime importance in reactions where the final product is temperature sensitive.[8]

Thermal effects can be used to explain several MW assisted phenomenon. Energy efficient heterogeneous catalysis (microwave assisted) is one of them. This efficiency can be attained by selectively maintaining a higher temperature of the catalyst than the bulk temperature of the solvent. Some authors have proposed the modification of the catalyst's electronic properties upon exposure to microwave irradiation[9] in order to explain the superior catalytic properties of catalysts under these conditions. However, other authors have reported that microwave irradiation has no effect on the reaction kinetics.[10]

Thermal effects can also be used to explain the lower yields from the oil bath experiments than those for the corresponding microwave-heated reactions. In the case of pure, microwave-transparent solvents, the added substances, either ionic or non-ionic, must contribute to the overall temperature profile when the reaction is carried out. It seems reasonable that when the substrates act as "molecular radiators" in channeling energy from microwave radiation to bulk heat, their reactivity might be enhanced. The concept and advantages of "molecular radiators" have been described by many authors.[11]

In the context of thermal effects of microwave synthesis, it is worthwhile to introduce the concept of a susceptor. A susceptor is an inert compound that efficiently absorbs microwave radiation and transfers the thermal energy to another compound that is a poor absorber of the radiation. Susceptors can profitably be used in catalysis and solvent-free green reactions. If the susceptor is a catalyst, the energy can be focused on the surface of the susceptor where the reaction takes place. In this way, thermal decomposition of sensitive compounds can be avoided. In contrast, transmission of the energy occurs through conventional mechanisms. In solvent-free or heterogeneous conditions graphite has been used as a susceptor. Ionic liquids have also been used as susceptors both in solution and under homogeneous conditions.[4]

1.2.2 Non-thermal Effects

All the properties of microwave reactions cannot be explained solely by thermal effects. In order to rationalize the effect of microwaves on the organic reactions, the concept of non-thermal or specific microwave effects has been floated by many researchers.

According to Miklavc[12] a large increase in the rates of chemical reactions occurs because of the effects of rotational excitation on collision geometry.

Non-thermal effects can be very well explained by keeping in mind each term of the Arrhenius law[13,6]

$$k = A \, \exp(-\Delta G^{*}/RT)$$

The pre-exponential factor, A, represents the probability of molecular impacts. The collision efficiency can be effectively influenced by mutual orientation of polar molecules involved in the reaction. Because this factor depends on the frequency of vibration of the atoms at the reaction interface, it could be postulated that the microwave field might affect this. A decrease in the activation energy ΔG^{*} could certainly be a major factor. Because of the contribution of enthalpy and entropy to its value $(\Delta G^{*} = \Delta H^{*} - T\Delta S^{*})$, it might be predicted that the magnitude of the $-T\Delta S^{*}$ term would increase in a microwave-induced reaction, because of greater randomness as a consequence of dipolar polarization.

On the basis of above criterion, multiple origins of specific microwave effects can be postulated. One of the major contributions comes from the **reaction media.**[14]

In case of **polar solvents**, either protic (*e.g.* alcohols) or aprotic (*e.g.* DMF, CH_3CN, DMSO *etc.*), there is a fair chance of interaction between microwaves and the solvent molecules. We can thus expect that the energetics of the reaction is governed by the energy transfer from the solvent molecules (present in large excess) to the reaction mixtures and the reactants. This mechanism is similar to that of conventional heating and it has been experimentally established that the rate of reaction in polar media is unaltered on moving from conventional to microwave heating.

Non polar solvents (*e.g.* xylene, toluene, carbon tetra-chloride, hydrocarbons) are transparent to microwaves, they therefore enable specific absorption by the reactants. When reactants are polar, energy transfer occurs from the reactants to the solvent and the results are different under the action of microwaves.[15]

Table 1.3 gives an overview of the absorbance capacities of commonly used solvents in the polymerization reactions.

The virtue of microwave heating is fully utilized in **solvent free** reactions. Microwaves in solvent free reactions not only lead to accelerated, economical and green reactions but also save one from the hassle of separation of products. The optimum use of microwaves can be accomplished by three methods.[16,17]

1. Reactions between the neat reagents in quasi-equivalent amounts, requiring, preferably, at least one liquid phase in heterogeneous media and leading to interfacial reactions.[18] Kinetic considerations for the reaction between two solids have been explained by considering the formation of a eutectic melt during the reaction.[19]
2. Solid–liquid phase-transfer catalysis (PTC) conditions for anionic reactions using the liquid electrophile as both reactant and organic phase and a catalytic amount of tetraalkylammonium salts as the transfer agent.[20]

Table 1.3 Classification of commonly used solvents based on their MW radiations absorbing capacity.

Absorbance capacity	Solvents
Low	Chloroform, dichloromethane, carbon tertrachloride,1,4-dioxane, THF, ethers, ethyl acetate, pyridine, triethylamine, toluene, benzene, chlorobenzene, xylene, hydrocarbons
Medium	Water, DMF, NMP, butanol, acetonitrile, HMPA, methyl ethyl ketone, nitromethane, o-dichlorobenzene, 1,2-dichloroethane, 2-methoxyethanol, acetic acid
High	DMSO, ethanol, methanol, nitrobenzene, formic acid, ethylene glycol

3. Reactions using impregnated reagents on solid mineral supports (aluminas, silicas, clays) in dry media.[17,21]

Reaction mechanism is a key determiner of the success of microwave application to any reaction. As microwave heating is associated with polarization of molecules, we can say that the efficacy of these syntheses depends on the alteration of polarity during the course of the reaction.

On going through the reaction profile, if stabilization of the transition state (TS) is more effective than that of the ground state (GS), this results in enhancement of reactivity as a result of a decrease in the activation energy (Figure 1.2), because of electrostatic (dipole–dipole type) interactions of polar molecules with the electric field. Reactions of this type include the following.

❖ Bimolecular reactions between neutral reactants, leading to charged products like amine or phosphine alkylation or addition to a carbonyl group.[5]

❖ Anionic bimolecular reactions involving neutral electrophiles These reactions comprise nucleophilic SN^2 substitutions, β-eliminations, and nucleophilic additions to carbonyl compounds or activated double bonds.[22]

❖ First order unimolecular reactions which involve development of dipolar intermediates. These dipolar intermediates increase the polarity from GS state to TS, thus bringing MW effect into the picture.

Several examples of **increased selectivity**,[23] in which the steric course and the chemo- or regio-selectivity of reactions can be altered under the action of microwave irradiation compared with conventional heating, have been observed. When competitive reactions are involved, the GS is common for both processes. The mechanism occurring *via* the more polar TS could, therefore, be favored under the action of microwave radiation.

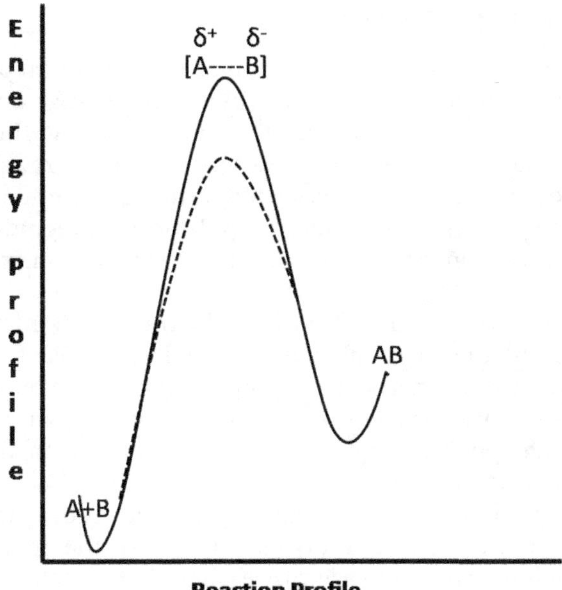

Figure 1.2 Reaction profile of conventionally heated reactions *vs.* microwave irradiated reactions (dotted line) with a polar transition state. Transition state is more stabilized on exposure to microwaves.

1.3   Loss Mechanisms

The two main loss mechanisms for non-magnetic materials are dielectric (dipolar) losses and conduction losses. Conduction losses dominate in metallic, high conductivity materials and dipolar losses dominate in dielectric insulators. Magnetic materials also exhibit conduction losses with additional magnetic losses such as hysteresis, domain wall resonance and electron spin resonance (FMR).

Loss tangent, which is in tangent form of loss angle, determines the ability of a substance to convert electromagnetic energy into heat.

$$\tan \delta = \varepsilon''/\varepsilon'$$

ε'' is called loss factor that refers to the efficiency of converting electromagnetic energy into the heat and ε' is called dielectric constant that indicates the ability of material to store electrical potential energy under applied electrical field. For effective microwave absorption a high loss tangent value is needed. When dielectric constant and loss tangent values of the solvents are close to each other, loss factor value becomes important to compare the abilities of different substances to convert electromagnetic energy into heat. Moreover, solvents that do not have dipole moment can be used in microwave ovens by adding polar additives like ionic liquids.[24]

1.4   Microwave Reactors

In order to fully understand the application of microwaves in polymerization and other organic syntheses, it is worthwhile to familiarize oneself with the instrumentation and techniques used. Though it is beyond the scope of this book to have an elaborate and detailed description of microwave design, the components are summarized in this segment. In a generic sense, microwave reactors comprise three components *viz.* **applicator, waveguide and cavity.**[24] Vacuum tubes and magnetrons are common sources of microwaves in the microwave reactors.

A microwave **applicator** is a device where the transfer of microwave energy from the source to the material being treated takes place. We can thus conclude that the more efficient the applicator is, the better is the reactor. The applicators can be modeled into a wide variety, depending on the reagent's packaging (powder, liquids, pellets) coupled with their dielectric characteristics and quantity to be heated.[25]

High power microwaves are generated by vacuum tubes. The magnetron and klystron are the most commonly used tubes for the generation of continuous wave power for microwave processing. At frequencies higher than 3 GHz, transmission of electromagnetic waves along transmission lines and cables becomes difficult, mainly because of the losses that occur both in the solid dielectric needed to support the conductor and in the conductors themselves. In order to overcome these losses hollow metallic tubes of uniform cross-section called **waveguides** are used for transmitting electromagnetic waves by successive reflections from the inner walls of the tube.[26]

Usually, resonant **cavities** are used as applicators. When microwaves traveling along a waveguide encounter an object (commonly referred to as a termination), a reflected wave travels back towards the source. Excessive reflected energy poses a threat to the magnetron, it is hence advisable that the resonant frequency of the loaded oven (and not the empty oven) should be close to the frequency of the magnetron. That is the reason why it is not advised to run empty domestic ovens. However, most commercial ovens are protected by a thermal automatic cutoff in case of poor matching between magnetron and oven.[24]

1.4.1   Single Mode and Multi-mode Instruments

Initially, domestic microwave ovens were used in the laboratory for synthesis purpose. However, with time the popularity of microwaves gained momentum and the synthetic chemists were able to outline their specific requirements and expectations for microwave reactors. This led to the development of a plethora of instruments, which can be broadly classified as:

- ❖ Single-mode apparatus
- ❖ Multi-mode apparatus

1.4.1.1 Single Mode Apparatus

A monomodal microwave device creates a standing wave pattern in the resonating cavity. This is generated by the interference of fields that have the same amplitude but different oscillating directions. The interface results in an array of nodes where microwave energy intensity is zero, and an array of antinodes where the magnitude of microwave energy is at its highest (Figure 1.3).

The design of a single mode reactor should be such that the sample encounters the antinodes of the standing wave pattern. One of the major limitations of single-mode apparatus is that only one vessel can be irradiated at a time. After the completion of the reaction period, the reaction mixture is cooled by using compressed air. Single mode reactors are simple to operate. They can process volumes ranging from 0.2 to about 50 ml under sealed-vessel conditions (250 °C, *ca.* 20 bar), and volumes around 150 ml under open-vessel reflux conditions.[24] Single-mode microwave reactors are generally used for small-scale drug discovery, automation and combinatorial chemical applications. An out and out advantage of single mode reactors is their high rate of heating, as the sample is always placed at the antinodes of the field, where the intensity of microwave radiation is the highest. In contrast, the heating effect is averaged out in a multi-mode apparatus.

1.4.1.2 Multi-mode Apparatus

In a multi-mode microwave reactor the radiation created by the magnetron is directly sent to the reaction cavity, where it is dispersed, thus avoiding the formation of a standing wave. In a multi-modal cavity, several samples can

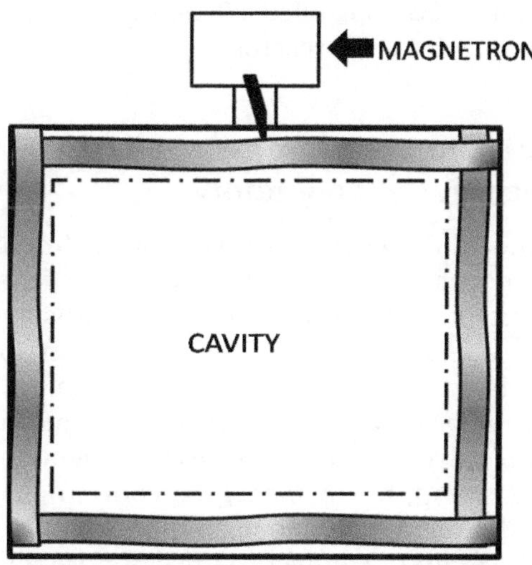

Figure 1.3 Schematic representation of a single mode microwave reactor.

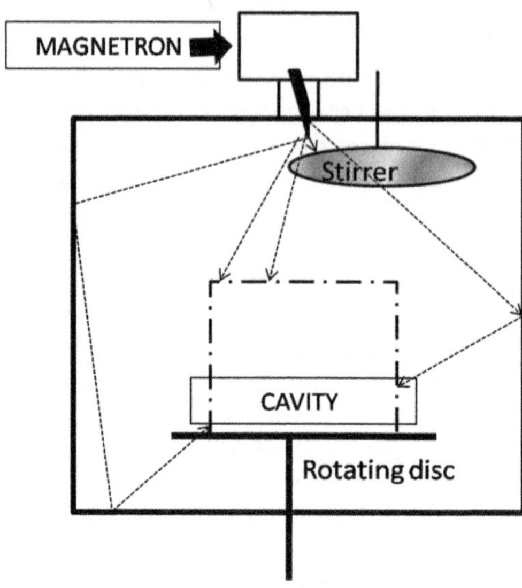

Figure 1.4 Schematic representation of a multi-mode microwave reactor.

be irradiated simultaneously. The domestic microwave oven is an example of this type of reaction assembly (Figure 1.4).

A multi-mode heating apparatus is used for bulk heating and carrying out chemical analysis processes such as ashing, extraction, *etc.* In large multi-mode apparatus, several liters of reaction mixture can be processed in both open and closed-vessel conditions. Recent research has resulted in the development of continuous-flow reactors for single- and multi-mode cavities that enable preparation of materials in kilograms.[27] A major limitation of multi-mode apparatus is that even with radiation distributed around them, heating samples cannot be controlled efficiently and a risk of hazardous explosion is associated with such reactors.

1.4.2 Domestic Microwave Ovens: Applications and Limitations in the Laboratory

Even today, multi-modal domestic microwave ovens find a good use in the laboratory as they are easy to procure and operate. Microwave safe vessels made from Teflon and polyetherimide, which can withstand pressures up to 80 atm and temperatures up to 250 °C, are available in abundance.[28] However, they are associated with several limitations, safety being a primary concern. One of the major issues with these ovens is the danger of explosion while heating organic solvents in an open vessel. Several modifications have been applied to this set-up, for instance, the conventional chemical reflux system could be used if the water condenser is outside the microwave cavity. In this case, it is necessary to connect the reaction vessel to the condenser through a port that ensures microwave leakage to safe limit.[29,30]

The problem of temperature measurement is another limitation. Classical temperature sensors fail to work when strong electric currents induced inside the metallic wires interfere with their operations. In order to overcome this, optical fiber thermometers are used. However, measurements are limited below 250 °C. For higher values, surface temperature infra-red cameras or pyrometers are used.[31] However, due to the volumic character of microwave heating, surface temperatures are often inferior to core temperatures.

1.5　Microwaves in the Laboratory: Hazards and Safety Measures

In the context of application in the laboratory, the use of microwave ovens for simple heating or defrosting in laboratories can pose a number of hazards, which include the following.[32–34]

- ➤ Ignition of flammable vapors.
- ➤ Exposure to microwave radiation from a faulty or modified unit.
- ➤ Electric shock from ungrounded or faulty units.
- ➤ Ignition of materials being heated.
- ➤ Pressure build-up in sealed containers.
- ➤ Sudden boiling of liquid in an open container following removal from an oven.
- ➤ Contamination of food products with chemical residues.

In order to minimize the risk of these hazards, some dos and don'ts are documented below:

Do not:

- ✗ Attempt to heat flammable liquids or solids, hazardous substances or radioactive materials in any type of microwave oven, whether domestic or laboratory-grade.
- ✗ Attempt to defeat the interlock switches that prevent a microwave oven from operating with the door open.
- ✗ Place any wires, cables, tubing *etc.* between the door and the seal.
- ✗ Modify in any way the mechanical or electrical systems of a microwave oven.
- ✗ Carry out unauthorized repairs on a microwave oven. Where a unit is suspected to be faulty, it should be disconnected from the power supply, removed from service and labeled with an appropriate tag while awaiting repair or disposal. Any irreparable or redundant microwave oven should be rendered inoperable by removal of the plug and cord, before disposal.
- ✗ Use a microwave oven in a laboratory for food preparation (or *vice versa*).
- ✗ Heat sealed containers in a microwave oven. Even a loosened cap or lid poses a significant risk since microwave ovens can heat material so

quickly that the lid can seat upward against the threads and containers can explode either in the oven or shortly after removal.

✗ Use bottles with a restricted neck opening (*e.g.* medical flats).

✗ Place metal objects of any kind in a microwave oven. This includes aluminum foil and plastic coated magnetic stirrer bars.

✗ Overheat liquids in a microwave oven. It is possible to raise water to a temperature greater than normal boiling point; when this occurs, any disturbance to the liquid can trigger violent boiling that could result in severe burns.

Do:

✓ Ensure that the oven cavity is adequately ventilated. The unit should be located on a clear, open bench and not in a location where the vents could be obstructed by books or equipment.

✓ Conduct regular inspections to ensure that the sealing surfaces are clean and do not show any sign of damage. The presence of arcing or burn marks may be indicative of microwave leakage.

✓ Ensure that microwave ovens are electrically grounded and connected using a properly rated three-pin cord and plug. As with all new laboratory equipment, microwave ovens should be inspected in accordance with the university's policy for electrical equipment to ensure compliance with this requirement.

✓ Report defects in equipment or difficulties in operation with a microwave oven promptly to the laboratory manager or supervisor.

✓ Where possible use microwave grade plastic vessels with a pressure relief valve. Where glass vessels are used, check them for cracks and flaws before using in the microwave.

✓ Use appropriate protective equipment when removing heated liquids from the oven.

1.6 Some Commercial Laboratory Products

With the advent and popularization of green chemistry, the development of safe, controllable and efficient microwave reactors has gained momentum. Slowly and steadily domestic microwaves in the laboratory are being replaced by specialized instruments. This part of the chapter outlines the development of commercial microwave reactors.

1.6.1 The Prolabo Products

❖ First Commercial Equipment, **Synthewave 402 and 1000** (Figure 1.5), for chemical synthesis under microwave, developed jointly by the Laboratory and Society Prolabo, from a device for Kjeldahl mineralization (Maxidigest TM MX 350).

❖ Use a closed rectangular waveguide section as cavity.

Figure 1.5 Synthewave 402.

Figure 1.6 Soxwave 100.

- ❖ Temperature measurement by infra-red surface.
- ❖ Temperature control by power modulation between 15 and 300 W.
- ❖ Possibility to work at reduced pressure or solvent reflux, mechanical stirring.
- ❖ Wide range in the amount of reagents used (0.2 g to 50 g) according to the size of the reactor.
- ❖ Possibility of a rise in level (up to 1 kg) on the Synthewave TM 1000.
- ❖ The Soxwave 100 (Figure 1.6) is a variant which has been designed for extractions such as Soxhlet, which are laborious.

❖ The extraction tube is capped with a cooling column.

❖ Optional temperature control during extraction was available.

❖ Maxidigest (one microwave unit 15 W to 250 W) and Microdigest (3–6 independent digestions at one time with integrated magnetic stirrers) are other variants for digestion.[35–37]

1.6.2   The CEM Products

❖ The best-selling **MARS System** (Figure 1.7) features a large multi-mode cavity, which allows for synthesis at scales up to a 5 L round-bottom flask or parallel synthesis using a multi-vessel turntable.

❖ The magnitude of microwave power available is 300 W. The optical temperature sensor is immersed in the reaction vessel for quick response up to 250 °C.

❖ Provided with a ground choke to prevent microwave leakage.

❖ The **Discover Series** of single-mode microwave synthesizers is available with a variety of options and accessories, including automation. The synthesizers are used for research scale reactions with volumes up to 75 ml.

❖ The applicator is constituted with two concentric cavities with aperture ensuring the coupling.

❖ It can operate at atmospheric conditions using open vessels and standard glassware (1 ml to 125 ml) or at elevated pressure and temperature using sealed vessels (0.5 ml to 10 ml sealed with a septum.[35–40]

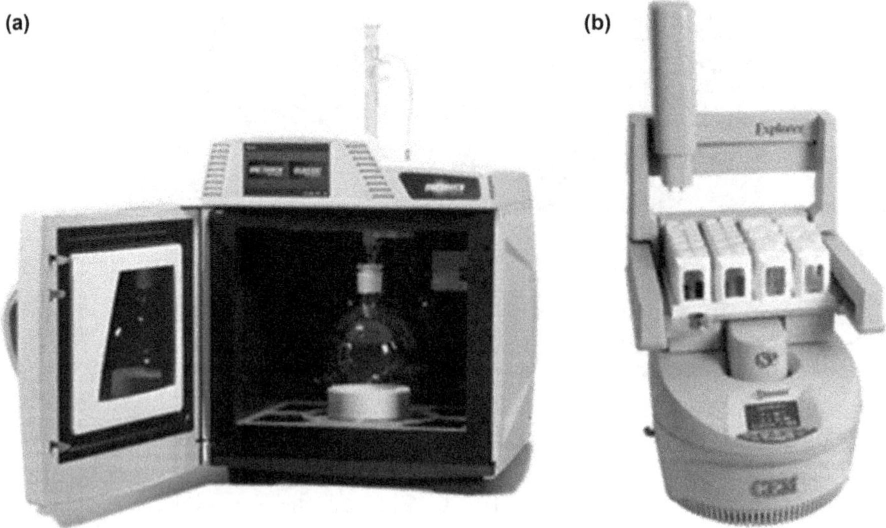

Figure 1.7   MARS System.

1.6.3   The Milestone Products

- ❖ The **ETHOS MR** (Figure 1.8) is constituted of a multi-mode cavity.
- ❖ Standard glassware or glass (420 ml up to 2.5 bar) and polymer reactors (375 ml up to 200 °C and 30 bar) with magnetic stirring can be used.
- ❖ The magnitude of microwave power available is 1 kW.
- ❖ The optical temperature sensor is immersed in the reaction vessel for quick response up to 250 °C.
- ❖ An infra-red sensor is also available.
- ❖ The **ETHOS CFR** is a continuous flow variant of the ETHOS MR.[41,42]

1.6.4   The Biotage (Formerly Personal Chemistry) Products

- ❖ The Smith Synthesizer (Figure 1.9) and Smith Creator are major products.[43–45]
- ❖ They are constituted by a closed rectangular waveguide section playing the role of cavity.
- ❖ Pressure and temperature sensors allow real-time monitoring and control of operating conditions.

1.6.5   The Plazmatronika Products

- ❖ Plazmatronika microwave reactors (Figure 1.10) are multi-modal microwave devices equipped with magnetic stirrer.
- ❖ Infra-red thermography is used to measure temperature.[46]

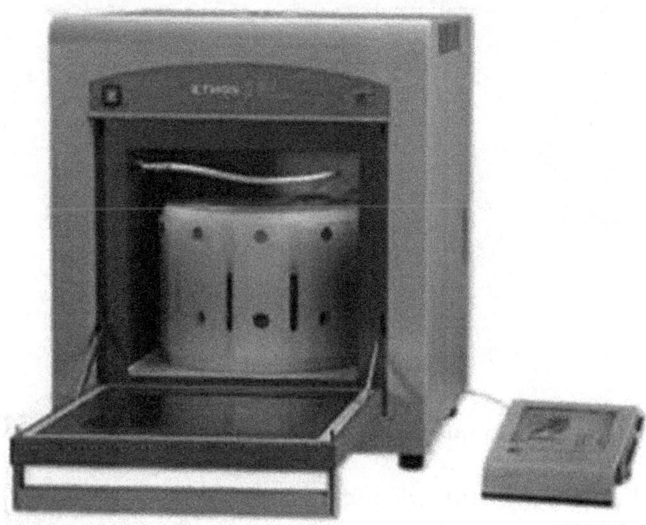

Figure 1.8   ETHOS MR.

Figure 1.9   Smith Synthesizer.

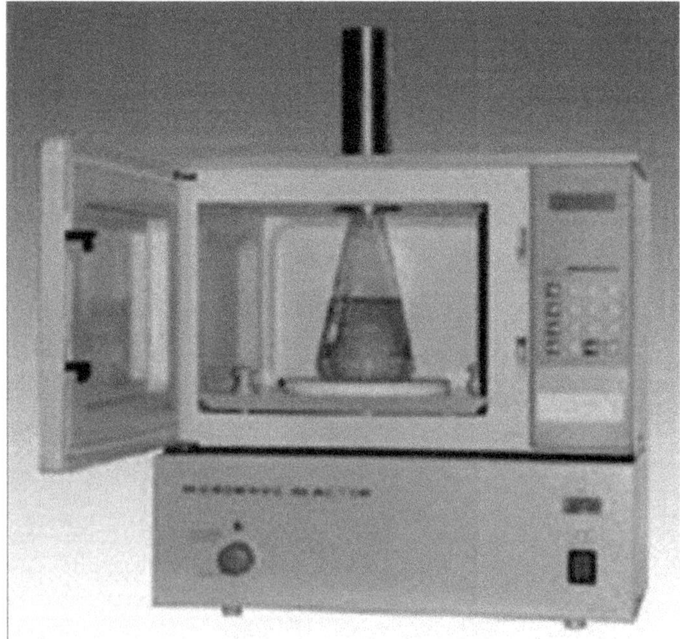

Figure 1.10   Plazmatronika microwave reactors.

Besides the aforementioned products, several research groups have modified and customized several microwave reactors as per their synthetic requirements. The major customizations include coupling with HPLC systems, UV systems, ultrasound deivces, change in reactor length, pressure and temperature sustatinabilty *etc.* Despite tremendous efforts in automation and in the development of new chemical methods in the past few years, there is still an overall deficiency in new chemical technology. With the ever-increasing thrust on novel, efficient and green methods of synthesis, it seems to be only a question of acquiring sufficient manpower and budget to achieve highly specific and modernized reactors in a reasonable time.[47]

References

1. http://en.wikipedia.org/wiki/Microwave.
2. http://en.wikipedia.org/wiki/Electromagnetic_spectrum.
3. P. Lidstrom, J. Tierney, B. Wathey and J. Westman, *Tetrahedron*, 2001, **57**, 9225.
4. M. A. Herrero, J. M. Kremsner and C. O. Kappe, *J. Org. Chem.*, 2008, **73**, 36.
5. L. Perreux and A. Loupy, *Tetrahedron*, 2001, **57**, 9199.
6. A. de la Hoz, A. Diaz-Ortiz and A. Moreno, *Chem. Soc. Rev.*, 2005, **34**, 164.
7. M. Lared, C. Moberg and A. Hallberg, *Acc. Chem. Res.*, 2002, **35**, 717.
8. P. Klán, J. Literák and S. Relich, *J. Photochem. Photobiol., A*, 2001, **143**, 49.
9. X. Zhang, D. O. Hayward and D. M. P. Mingos, *Chem. Commun.*, 1999, 975.
10. D. Bogdal, M. Lukasiewicz, J. Pielichowski, A. Miciak and Sz. Bednarz, *Tetrahedron*, 2003, **59**, 6488.
11. M. Lukasiewicz, D. Bogdal and J. Pielichowskia, *Adv. Synth. Catal.*, 2003, **345**, 1269.
12. A. Miklavc, *ChemPhysChem*, 2001, **2**, 552.
13. J. G. P. Binner, N. A. Hassine and T. E. Cross, *J. Mat. Sci.*, 1999, **30**, 5389.
14. W. C. Shieh, S. Dell and O. Repic, *Org. Lett.*, 2001, **3**, 4279.
15. M. Nuchter, B. Ondruschka, A. Jungnickel and U. Muller, *J. Phys. Org. Chem.*, 2000, **13**, 579.
16. A. Loupy, *Top. Curr. Chem.*, 1999, **206**, 155.
17. C. R. Strauss and R. W. Trainor, *Aust. J. Chem.*, 1995, **48**, 1665.
18. C. R. Strauss, *Aust. J. Chem.*, 1999, **52**, 83.
19. K. Tanaka and F. Toda, *Chem. Rev.*, 2000, **100**, 1025.
20. R. S. Varma, *Green Chem.*, 1999, **1**, 43.
21. S. Caddick, *Tetrahedron*, 1995, **51**, 10403.
22. M. Szware, *Ion Pairs in Organic Reactions*, Wiley Interscience, New York, 1972/1974, vol. 1 & 2, pp. 21–23.
23. A. Loupy, P. Pigeon, M. Ramdani and P. Jacqualt, *Synth. Commun.*, 1994, **24**, 159.
24. O. C. Kappe, *Angew. Chem., Int. Ed.*, 2004, **43**, 6250.
25. M. Mladenović, A. Marinčić and B. Milovanović, *J. Microwave Power EE*, 1989, **24**, 166.

26. T. V. C. T. Chan and H. C. Reader, *Understanding Microwave Heating Cavities*, Artech House Publishers, Boston and London, 2000.
27. www.rsc.org/images/evaluserve_tcm18-16758.pdf.
28. A. Herzog, R. Klingner, U. Vogt and T. Graule, *J. Am. Ceram. Soc.*, 2004, **87**, 784.
29. H. Sieber, C. Hoffmann, A. Kaindl and P. Greil, *Adv. Eng. Mater.*, 2000, **2**, 105.
30. G. C.-T. Wei, *Am. Ceram. Soc. Comm.*, 1983, **66**, C-111.
31. I. S. Hukvari, H. J. Albert, V. G. Moon and M. R. Steffenson, *U. S. Pat.* 4882128 A, 1989.
32. www.anton-paar.com.
33. J. M. Kremser and C. O. Kappe, *Eur. J. Org.Chem*, 2005, 3672–3679.
34. http://www.nottingham.ac.uk/safety/documents/microwaves.pdf.
35. C. Limousin, J. Cléophax, A. Petit, A. Loupy and G. Lukacs, *J. Carbohydr.Chem.*, 1997, **16**, 327.
36. A. Loupy, A. Petit, J. Hamelin, F. Texier-Boullet, P. Jacquault and D. Mathe, *Synthesis*, 1998, 1213.
37. CEM Corporation, Matthews (NC), www.cemsynthesis.com.
38. D. P. Manchester, W. P. Hargett and E. D. Neas, *EU Pat.* 0455513, 1991.
39. B. W. Renoe, E. E. King and D. A. Yurkowich, *EU Pat.* 0604970, 1994.
40. W. E. Jennings, D. P. Manchester, E. E. King and D. A. Barclay, *U. S. Pat.* 5796080, 1998.
41. L. Favretto, *Mol. Diversity*, 2003, **7**, 287.
42. http://www.milestonesci.com/ethos-up.html.
43. P. Tierney and P. Lidstrom, *Microwave Assisted Organic Synthesis*, Blackwell Publication, Oxford, UK, 2005.
44. C. O. Kappe and A. Stadler, *Microwaves in Organic and Medicinal Chemistry* , Wiley-VCH, Weinheim, 2005.
45. www.biotage.com.
46. http://www.plazmatronika.pl.
47. *Microwaves in Organic Synthesis*, ed. A. Loupy, Wiley-VCH Verlag GmbH & Co. KGaA, Weinheim, 2006.

CHAPTER 2

Radical Polymerization

2.1 Introduction

The polymerization method, by which a polymer is formed by the successive addition of free radical monomer unit is known as free radical polymerization. Free radicals can be formed by a number of different mechanisms usually involving separate initiator molecules. The generation of initiating free radicals in turn leads to its addition to the non-radical monomer unit(s) to form radical monomer units, thereby growing the polymer chain.

Free radical polymerization is a key synthesis route for obtaining a wide variety of different polymers and material composites. The relatively non-specific nature of free radical chemical interactions makes this one of the most versatile forms of polymerization available and allows facile reactions of polymeric free radical chain ends and other chemicals or substrates. In 2001, more than one third of the total polymers produced in the United States were produced by free radical polymerization.[1,2]

The popularity of free radical polymerization in industry and research laboratories is due to the combined virtues of versatility, synthetic ease and compatibility with a wide variety of functional groups, coupled with its tolerance to water and protic media. These features lead to the development of emulsion and suspension techniques, which greatly simplifies the experimental setup and hence commercial adoption.[3] Vinylic polymers such as polyethylene, poly(methyl methacrylate), poly(tetrafluoroethylene), poly(vinylchloride), poly(vinyl acetate), poly(acrylonitrile) and polystyrene are produced by radical polymerization.[4] Microwave-assisted free radical polymerization was first reported in 1979, followed by the report on polymerization of hydroxyethyl methacrylate (HEMA) under MW without any radical initiator. It was demonstrated in these studies that the temperature of the reaction mixture for both a microwave-irradiated polymerization and a thermal polymerization gave similar profiles, although the

RSC Green Chemistry No. 35
Microwave-Assisted Polymerization
By Anuradha Mishra, Tanvi Vats and James H. Clark
© Anuradha Mishra, Tanvi Vats and James H. Clark 2016
Published by the Royal Society of Chemistry, www.rsc.org

microwave-assisted polymerization was significantly faster.[5] The accelerated polymerization of methyl methacrylate (MMA), methyl acrylate (MA), and styrene under microwave irradiation has also been reported at the same time.[1]

The main drawback of this otherwise widespread technique is poor control of molecular weight and polydispersity indices. In order to overcome these shortcomings, controlled free radical techniques were developed over the years.

This chapter deals with microwave-assisted free radical homopolymerization, copolymerization, graft copolymerization, controlled radical polymerization, emulsion polymerization and their merits over conventional heating techniques.

2.2 Free Radical Homopolymerization

Microwave-assisted free radical polymerization has been receiving ever increasing attention mainly due to highly accelerated reaction rates. Not only this, microwave-assisted reactions are affordable, more controllable and give uniformly processed products. Microwave-assisted free radical polymerizations were initially investigated by Gourdenne *et al.* in 1979.[5] They performed the cross-linking of an unsaturated polyester with styrene under microwave irradiation. The same group demonstrated that the use of microwave radiations in polymerization of hydroxyethyl methacrylate eliminated the necessity of a radical initiator.[6] The reaction involved treatment of 2-Hydroxyethylmethacrylate (HEMA) in bulk (without any radical initiator) with a polarized microwave beam (2.45 GHz) inside a wave guide at various electrical powers (P_0). A significant tell-tale sign is the polymerization of the fluid monomer to form a solid material. The authors described the various steps of the reaction by profiling the variation with time of the temperature of the sample and of the part (P_u) of the electrical power degraded in the polymerizable medium because of dielectric loss. They reported that, though the temperature of the reaction mixture for both a microwave-irradiated polymerization and a thermal polymerization gave similar reaction profiles, the microwave-assisted polymerization gave significantly increased reaction rates.

MW-assisted polymerization of MMA has been studied by various research groups. Madras and Karmore gave an elaborate account of kinetics of polymerization and simultaneous depolymerization of MMA under microwave irradiation, in order to fully comprehend the stability and characterization of the resulting polymer.[7] They proceeded by varying the initiator and monomer concentrations, while using distribution kinetics to study the change in molecular weight distribution with time. The authors claimed that an equilibrium between polymerization and depolymerization was reached within 10 min (with a polydispersity of two), resulting in equal polymer distributions for different initiator and monomer concentrations. The MW-assisted bulk polymerization of MMA in the presence of benzoyl peroxide as the initiator was studied by Jovanovic *et al.*[8] in a comparative account with experiments under conventional heating at different temperatures; they reported an acceleration of the reaction rate because of a reduction of the activation energy in polymerizations under MW irradiation. Boey and co-workers reported an increase in polymerization rates for the MW-assisted polymerizations of MMA, styrene and methyl acrylate (MA).[9,10] They performed the polymerization reaction at three different MW powers, namely, 200, 300, and 500 W and calculated the percentage conversion of the reaction by Fourier transform infrared (FTIR) spectroscopy. The results indicate that a similar comparable temperature of about 52 °C was found for all the microwave power settings tested. They compared the MW polymerization process with the thermal method at 52($\pm$1) °C under comparable reaction conditions. The enhancement of the reaction rate of MW polymerization in comparison to the thermal method was found to be 275% for the 500 W, 220% for the 300 W, and 138% for the 200 W. The results indicated a significant correlation between the reaction rate enhancement and the level of microwave power used.

Ritter and co-workers reported the synthesis of (meth)acrylamides and the subsequent polymerization of those monomers in a monomodal microwave synthesizer.[11] The group was first to report the direct solvent free amidation of (meth)acrylic acid with aliphatic and aromatic amines.[12] The corresponding (meth)acrylamides were obtained sans coupling agents after 30 min of MW irradiation in excellent yields (close to 96%). Several experiments were directed towards the understanding of stereoselective synthesis of chiral (meth)acrylamides from (meth)acrylic acid and (*R*)-1-phenylethylamine under the influence of MW irradiation (Scheme 2.1).[13,14]

The MW-assisted synthesis of *N*-((*R*)-1-phenylethyl)(meth)acrylamide presented a contrasting study with respect to thermal conditions. A 15 min period of MW irradiation in a monomode reactor resulted in amide formation, while Michael addition was the preferred reaction under thermal heating. The selective amide formation in MW conditions was explained by the formation of highly polar, strongly absorbing intermediates zwitterions and salts. The MW-assisted amidation performed in the presence of 2,2-azoisobutyronitrile (AIBN) resulted in a single step synthesis of optically active polymers. Ritter and co-workers prepared an amide by the reaction of methacrylic acid and 3-(dimethylamino)-1-propylamine for 1 min of MW irradiation (150 °C). Lower critical solution temperature behavior in aqueous solution was observed on homopolymerization of the purified monomer in solution of toluene using AIBN as the initiator.[15] Singh *et al.* used a domestic MW oven to polymerize acrylamide in aqueous solution with potassium peroxodisulfate as the initiator.[16] The study explained the dependency of the monomer conversion on the MW power, irradiation time, initiator and monomer concentration. The high yielding (98.5% conversion), optimized conditions were found with 80% MW irradiation (98 °C) for 50 s with potassium peroxodisulfate concentration as 2×10^{-3} M and acrylamide concentration as 0.56 M. The study emphasized that the application of identical conditions of temperature and concentration to the experiment performed

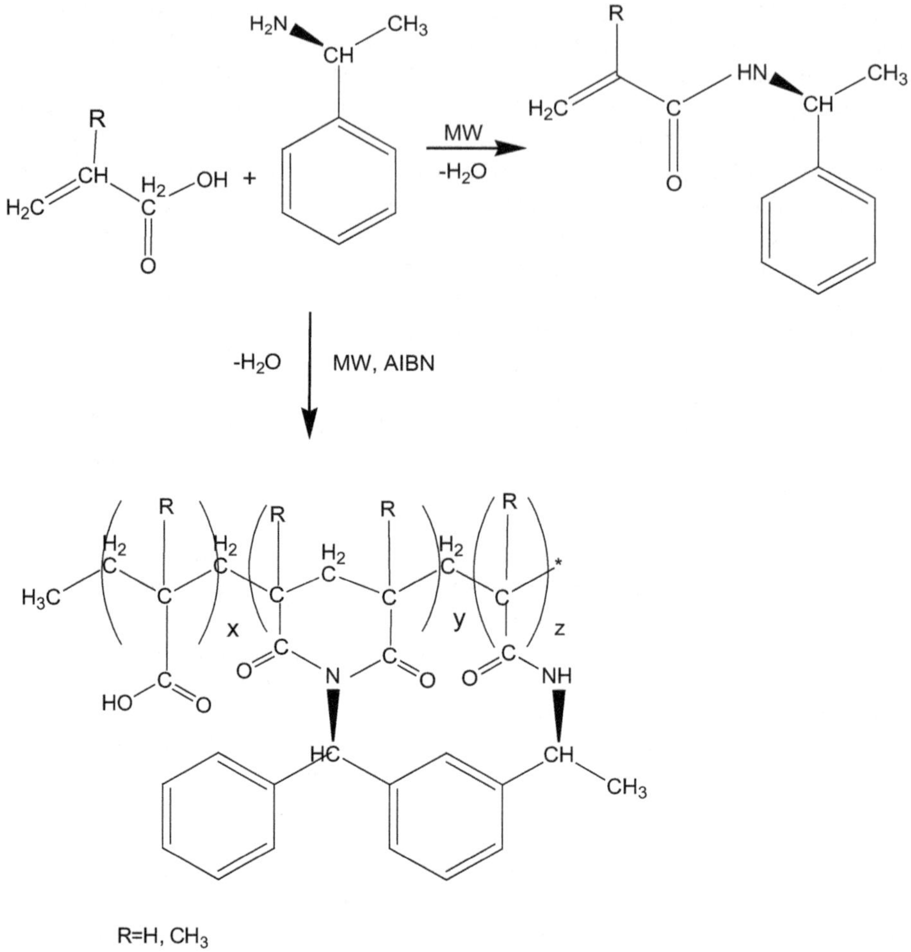

Scheme 2.1 MW-assisted synthesis of chiral (meth)acrylamides from (meth)acrylic acid.[13,14]

in a thermostatic water bath did not result in any polymerization. Fischer *et al.* investigated the MW-assisted chain transfer polymerization (telomerization) of poly-*N*-isopropylacrylamide (PNIPAM), poly-*N,N*-dimethylacrylamide (PNDMAM) and poly-*N*-{3-(dimethylamino)propyl}acrylamide (PN3DMAPAM) as well as of co-polymers of PNIPAM and PNDMAM.[17] The case study comprised comparative reactions in superheated yet subcritical methanol (80–170 °C) and under solvent-free conditions induced by MW irradiation. The reactions were compared in terms of product yield and quality to those obtained under standard reflux conditions. Though the MW irradiations dramatically reduced the reaction times, the average molar mass of the polymers dropped by 30% in these experiments. This was explained by an effect that is most likely caused by the higher polarity of the reaction mixture under solvent-free conditions. Lu and co-workers used a domestic microwave oven to investigate the influence of carriers (Al_2O_3, SiO_2, and MgO) on the MW-assisted polymerization of acrylamide and 2-ethylhexyl acrylate.[18] They concluded that the carrier quantity strongly influenced the polymer yield.

Sitaram and Stoffer studied the microwave-assisted polymerization of styrene with different radical initiators.[19,20] They irradiated the solutions of 5 mol percent initiator in styrene for two minutes in a domestic microwave oven at 800 W. Significant polymerization within two minutes only occurred with AIBN, *tert*-butyl peroxybenzoate and *tert*-amyl peroxybenzoate as initiators. Apart from decreased reaction times, the reactions under MW conditions attained the final temperature faster than the conventionally heated samples. However, in terms of conversions and the molecular weights of the polymers obtained, both the conditions gave comparable results. Polymerization of styrene was also studied by Boey and co-workers. However, in contrast to the above results, they reported an accelerated polymerization of styrene under microwave irradiation.[10,21] They observed a sharp and large auto-acceleration (Trommsdorf effect) under microwave irradiation, whereas only a gradual auto-acceleration effect was observed during conventional heating. This sharp auto-acceleration was also observed for the microwave polymerization of methyl methacrylate but not for methyl acrylate. This absence of a sharp transition for MA was attributed to chain transfer reactions that occur during the polymerization of MA.

Several monomer combinations for solid phase MW-assisted copolymerization has been tried by various researchers (Table 2.1). Polymerization of 4-nitrophenyl acrylate in a monomodal microwave reactor was studied by Bovin and co-workers.[22] MW irradiation accelerated the reactions as compared to conventional heating. In an interesting contrast, the polymers from the microwave-supported synthesis had significantly lower polydispersity indices. The observation implied that the termination of the polymer chains mainly proceeded by disproportionation reaction instead of recombination reaction (which is dominant in the case of thermal heating). They also emphasized on the reproducibility of the microwave-aided reactions.

Cortizo studied the influence of MW power and irradiation time on the yield and molecular weight of the polymerization of diisopropyl fumarate under MW irradiation (domestic MW oven).[23] Taking the study further, Alessandrini and co-workers performed polymerization of different dialkyl fumarates in a domestic MW oven using benzoyl peroxide as the initiator.[24]

Cai and co-workers investigated the MW-assisted, fullerene-initiated charge-transfer bulk polymerization of N-vinylcarbazole (N-VC) (Scheme 2.2c).[25] Fullerenes form $C_{60}^{+\bullet}$-N-VC$^{-\bullet}$ ion-radical pairs that initiate the polymerization of N-VC. The MW-assisted polymerization was much faster than the conventional heating methods. Interestingly, the authors attributed this to the higher temperature during the course of the microwave polymerizations. Mattos and co-workers reported the possibility of utilizing a domestic microwave oven to perform the free radical bulk polymerizations

Table 2.1 Common monomer combinations for soild-phase MW-assisted copolymerization.

 maleic anhydride	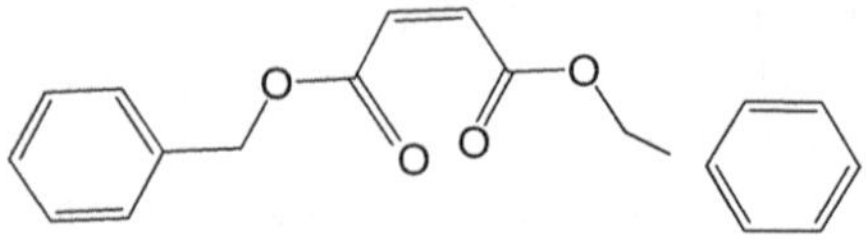 dibenzyl maleate
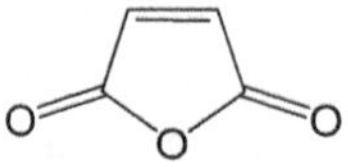 maleic anhydride	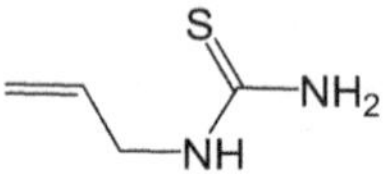 allylthiourea
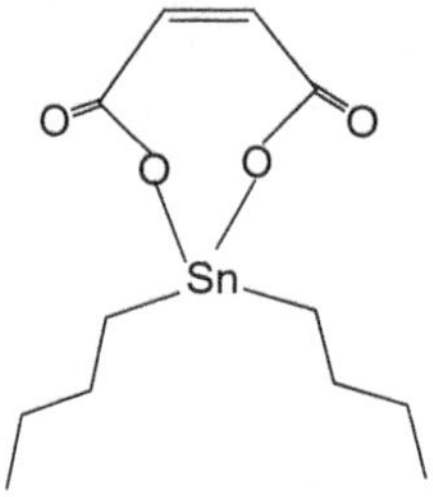 dibutyltin maleate	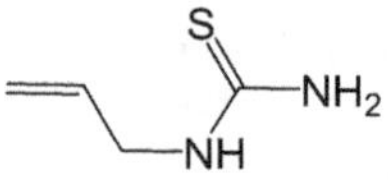 allylthiourea
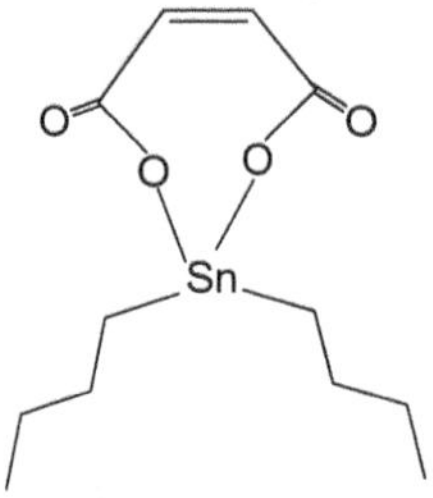 dibutyltin maleate	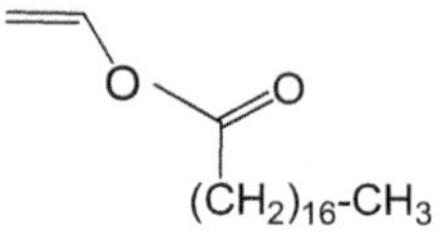 stearic acid viny lester

of vinyl acetate, styrene, methyl methacrylate and acrylonitrile.[26] The polymerizations reportedly proceeded at least 60 times faster (compared with conventional heating), utilizing (AIBN) as initiator. They also claimed that as the acrylonitrile can absorb MW irradiations, its polymerization could also be performed in the absence of AIBN. Elsabeé *et al.* reported the microwave-assisted homopolymerization of *N-p*-bromophenylmaleimide (BrPMI) using AIBN as initiator.[27] It is noteworthy that this polymerization could not be performed under conventional heating even at 135 °C with AIBN as initiator, but occurred readily within 10 min using microwave heating.

Biswal and co-workers reported free radical polymerization of acrylonitrile (AN) in the presence of Co(III) complex, [Co(III) en$_3$]Cl$_3$ and ammonium persulphate (APS) in domestic microwave ovens. Irradiation at low power and time produced more homogeneous polymers with high molecular weight and low polydispersity in comparison to the polymer formed by conventional heating method.[28]

2.3 Free Radical Copolymerization

A heteropolymer or a copolymer is a polymer derived from two or more monomeric species as opposed to one homopolymer. MW-assisted free radical copolymerization of styrene and MMA was investigated by Greiner and co-workers. They used a monomode MW reactor, with temperature control to devise a comparative study by using different organic peroxide initiators in a solution of toluene and *N,N*-dimethylformamide (DMF).[29] They found that the homopolymerizations of styrene in DMF gave a significantly higher monomer conversion under MW conditions. The polymerizations in toluene showed no influence under MW irradiation. For co-polymerization, the comonomer incorporation remained unaffected by the MW irradiation. In order to understand the finer points of the above reaction (Scheme 2.2), Stange and Greiner performed a further, more careful, comparative study on it.[30] From the results, it was demonstrated that MW conditions led to an enhanced monomer conversion in DMF with *tert*-butylperbenzoate, while all other copolymerizations remained unaffected. This solvent dependence was explained by the fact that DMF heats up much more quickly under MW irradiation than toluene. This in turn led to more radical formation during the early stage of the copolymerization (using DMF as solvent). In copolymerization experiments of styrene with MMA, and of butyl methacrylate with styrene and isoprene, a 1.7 fold increase in the reaction rate was observed by Fellows.[31] This increase was due to an increased radical concentration because of a rapid orientation of the AIBN fragments after decomposition.

Agarwal *et al.* used a monomode MW reactor to study copolymerization of 2,3,4,5,6-pentafluorostyrene and *N*-phenylmaleimide. They used anisole as a solvent and AIBN as the initiator, the temperature of the reaction was maintained at 70 °C.[32] The group observed an increase in the polymerization rate under MW conditions but lower limiting conversion compared with conventional thermal experiments. However, the copolymer composition remained unaffected by the mode of heating.

With the aid of a domestic MW oven, Lu *et al.* investigated the MW-assisted initiator-free copolymerization of 2-(dimethylamino)ethyl

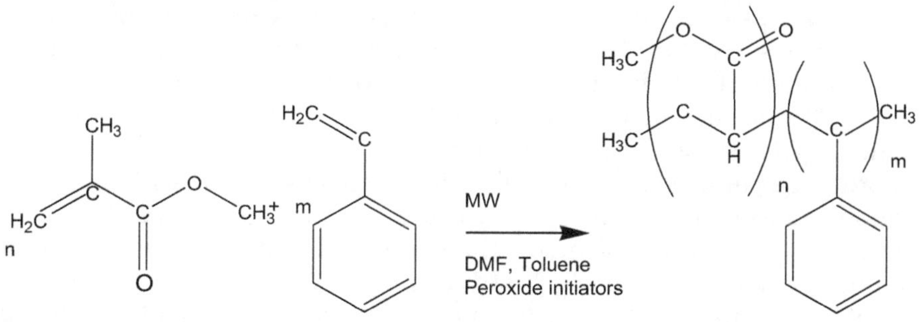

Scheme 2.2 Free radical copolymerization of styrene and MMA.[30]

methacrylate and 1-allylthiourea.[33] They also synthesized the copolymer–copper coordinates Cu-P(DM-*Co*-AT) by the coordination between Cu^{2+} and the copolymer. The study primarily aimed to understand the influence of irradiation time on monomer conversion (maximum attained was 92%) and irradiation power on the inherent viscosity (which was computed to 37.75 ml/g).

Rodriguez and co-workers reported random bulk copolymerizations of HEMA and MMA.[34,35] The results clearly demonstrated the accelerated rates on using MW irradiations. Under MW conditions polymerizations were completed within 45 min as compared to 125 min using conventional heating. Also, polymers synthesized under microwave irradiation showed higher molecular weights and lower PDI values than the thermally synthesized polymers (microwave irradiation: 1.36–2.08, conventional heating: 4.1). The microstructure and the physical properties of the polymers remained the same, irrespective of the mode of heating used. Lu *et al.* performed a series of solid-state random copolymerizations using various maleate derivatives.[36–39] They prepared a list of reactivity ratios for the different combinations of monomers. The group performed a 32 s MW irradiated reaction (at 45 °C) of the copolymerization of maleic anhydride with dibenzyl maleate.[36] The resulting copolymers were used for the preparation of superabsorbent oil resins. The copolymerization of ground maleic anhydride and allylthiourea led to the synthesis of water-soluble polymers with metal ion complexing abilities.[37] Water solubility and the metal complexing capacities of these copolymers were pH dependent. The copolymerization of dibutyltin maleate and allyl thiourea yielded heat-stabilizing organotin polymers.[38] The classic method of addition of a radical scavenger was applied to study the mechanism of the copolymerization. It was demonstrated that an increase in MW power led to an increase in radical concentration and that MW conditions yielded polymers, even in the absence of radical initiators. Additionally, a drastic increase in reaction speed from several hours down to a few minutes was observed for the solid-phase copolymerization of itaconic acid and acrylamide, sodium acrylate and *N*,*N*-methylenebiscrylamide, acrylamide and maleic acid anhydride, and also for the copolymerization of dibutyltin maleate and stearic acid vinyl ester.[39] In order to further improve the MW absorption, the effect of carriers was also studied.[38–41] Aluminum oxide was found to be a better carrier than silicon oxide as it enhanced the reaction speed of the copolymerization of dibutyltin maleate and allyl thiourea by a greater measure.[38,39] A more detailed study concludes that aluminum oxide being an alkaline carrier suited acidic or neutral monomer combinations, while silicon oxide was more suitable for alkaline systems as it is an acidic carrier.[40] In another mechanistic study performed by Lu and co-workers, it was observed that free radical polymerizations in the presence of silicon oxide or aluminum oxide proceeded by a radical mechanism, whereas in the presence of magnesium oxide the same polymerizations occurred by both radical and anionic mechanisms.[41]

2.3.1 Graft Copolymerization

Grafting is a chemical process by which polymer chains are attached to backbone polymer. With the aid of grafting synthetic polymers on natural polysaccharides,[42,43] we can obtain highly customizable matrices with hybrid properties suitable for different applications. Grafting of vinylic monomers on polysaccharides is a well-studied reaction. The three main strategies applied for grafting polymer chains on polysaccharides are: (i) grafting through, (ii) grafting on and (iii) grafting from.[44]

(i) **Grafting through**

A low molecular weight monomer is radically copolymerized with a (meth)acrylate functionalized macromonomer.

- The method permits incorporation of macromonomers that have been prepared by other controlled polymerization processes into a backbone prepared by a CRP.
- Branches can be distributed homogeneously or heterogeneously, depending on the reactivity ratio of the terminal functional group on the macromonomer and the low molecular weight monomer.

(ii) **Grafting from**

- The mechanism for "grafting from" polymerization includes incorporation of initiating site by copolymerization through radically transferable atoms along the polymer backbone.

(iii) **Grafting to**

- This is a very efficient method in which grafting takes place by various click chemistry reactions. The most prominent feature of these reactions is that they are site specific reactions.
- The versatility of this approach is demonstrated by the fact that it is applied to the preparation of well-defined star-shaped copolymers, loosely grafted copolymers and several densely grafted structures.[45]

The efficacy of application of MW irradiation to graft copolymerization has been very well studied and documented. These reactions are, on the one hand, safe and convenient to operate; on the other, they limit the usage of chemicals and shorten reaction time from hours to minutes to even seconds with high batch-to-batch consistency.[46–53] Besides these advantages, polysaccharide grafting under MW irradiation can be performed in open reaction vessels with reduced solvent content, enabling the reactions to go to completion, unlike conventional thermal grafting reactions in which an inert atmosphere is often required.[54]

Mainly three types of MW grafting reactions have been described by scientists; (i) in solution under completely homogeneous conditions where all the reaction contents are fully miscible with no phase separation; (ii) in suspension where the reactants are not fully miscible; either polysaccharide and/or monomer/catalyst are immiscible; (iii) in solid phase where the polysaccharide monomers and initiator are impregnated on neutral solid support.[54]

Singh *et al.*, studied the MW-assisted polymerization of MMA,[55] acryla-mide[56] and acrylonitrile[57] onto chitosan (Scheme 2.3). They performed a comparative study and observed that the experiments using a redox initiator had greater yields and higher grafting efficiencies under MW irradiation as compared to the conventional thermal heating. Even an initiator-free poly-merization was observed to be successful under MW conditions. Similar results were obtained when MW-assisted grafting was performed with acrylamide[46] and acrylonitrile[58] onto guar gum, acrylonitrile[59] onto cassia siamea seed gum and acrylamide[60] onto potato starch. Using a very low concentration of potassium persulfate as initiator, acrylamide could be efficiently grafted onto potato starch under microwave irradiation, and for the grafting O_2 removal from the reaction vessel was not required.[60] Sid-dhanta and co-workers used a domestic MW oven to study grafting of MMA onto κ-carrageenan.[61] The influence of irradiation time, MMA/κ-carrageenan ratio, reaction temperature and initiator concentration on MMA conversion and grafting efficiency was investigated. A graft copolymer hydrogel of κ-carrageenan and acrylamide has been synthesized in aqueous medium at pH 7 in the presence of the initiator potassium persulfate, using microwave irradiation.[62]

Sodium acrylate was polymerized under MW conditions (in a domestic MW oven), with potassium persulfate in the presence of corn starch and poly(ethyleneglycol) diacrylate as cross-linker, by Peng and co-workers.[63]

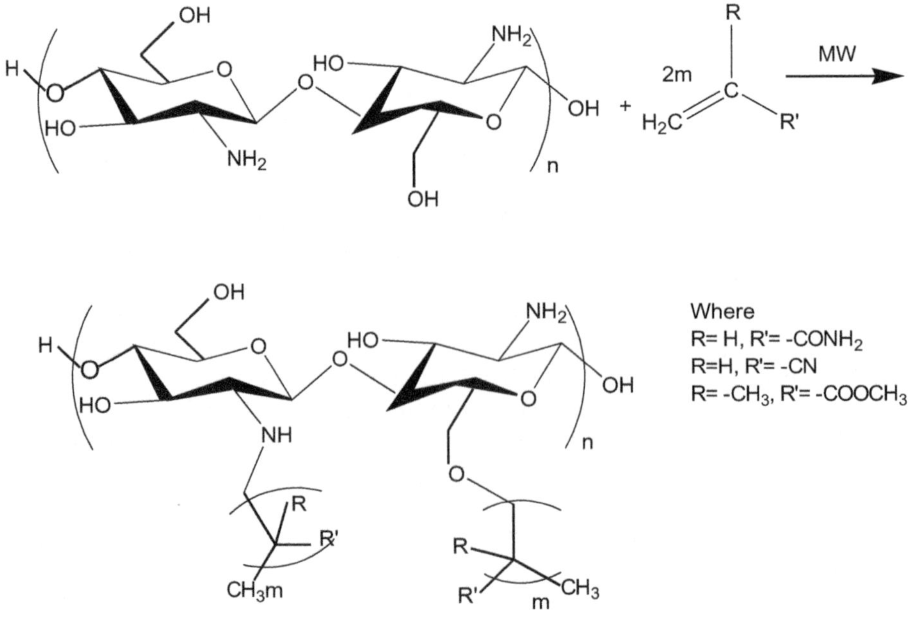

Scheme 2.3 MW-assisted polymerization of MMA, acrylamide, and acrylonitrile onto chitosan.[55–57]

It was found that microwave irradiation could substantially accelerate the synthesis, without the need to remove O_2 or the inhibitor. They employed an orthogonal test of $L_9(3^4)$, designed to study the effects of relevant factors on absorbency and yield of products, including microwave power, irradiation time, initiator amount and crosslinker content. From the optimized experimental results, it was observed that 10 min of microwave irradiation at 85–90 W could produce a cornstarch-based superabsorbent with a swelling ratio of 520–620 g/g in distilled water and solubility of 8.5–9.5 wt %.

Wang and co-workers performed the MW-assisted free radical crosslinking of acrylic acid and 2-(acrylamido)-2-methylpropane-1-sulfonic acid with *N*,*N*-methylenebisacrylamide in the presence of corn starch.[64] In terms of water absorbing properties, the authors showed that the swelling behavior of a material obtained after MW irradiation was improved as compared to that of a material synthesized by the conventional method. This was attributed to the occurrence of well-dispersed pores in the resin after MW-assisted synthesis, as a result of the efficient homogeneous heating.

Pawluczyk *et al.* used a multi-mode MW reactor to explain the scale-up of the MW-assisted living free radical polymerization of styrene from Rasta resin beads.[65] The authors found that this heterogeneous reaction was not directly scalable from reaction conditions determined on a lower scale. After optimization, however, 40–100 g quantities of Rasta resins with loading levels >3.8 mmol g^{-1} could be achieved.

Zhang and co-workers studied graft copolymerization of artemisia seed gum with acrylic acid under microwave and its water absorbency.[66] Singh and co-workers presented neutral alumina supported synthesis of cassia marginata gum-*g*-poly(acrylonitrile) and synthesis of graft copolymers of poly(acrylonitrile) and xyloglucan under microwave irradiation in dry medium. The authors optimized grafting conditions and characterized the representative graft copolymer by Fourier transform-infrared, Thermo gravimetric analysis, Scanning electron microscopy, and X-Ray Diffraction (XRD) measurement, taking *Cassia marginata* gum (CM) as reference. On comparison with grafting done in aqueous medium (microwave promoted and conventional), the authors observed that much higher % grafting and % efficiency could be obtained in the microwave promoted alumina supported grafting in dry medium.[67] Synthesis of graft copolymers of methyl methacrylate onto saccharum *spontaneum* fibers under the influence of microwave radiation was carried out. Graft copolymers have been found to be more moisture resistant and also showed higher chemical and thermal resistance than their conventionally prepared counterparts.[68,69] Ahuja and co-workers synthesized Xanthan-*g*-poly(acrylamide) by two methods, namely MW-assisted and ceric-induced graft copolymerization, characterized by FT-IR, DSC, XRD and SEM studies. They formulated matrix tablets of diclofenac sodium using graft copolymer as the matrix by direct compression technique. Release behavior of the graft copolymer was evaluated using USP type-II dissolution apparatus in 900 ml of phosphate buffer (pH 6.8), maintained at 37 °C and at 50 rpm. It was observed that MW-assisted grafting provided graft copolymer with higher % grafting in a shorter time in comparison to the ceric-induced grafting. Besides, the % grafting was directly proportional to the power of microwave and/or time of exposure. The drug release proceeded by zero-order kinetics, and the faster release of drug was observed from the graft copolymer matrix as compared to the xanthan gum matrix. It was observed that grafting reduces the swelling, but increases the erosion of xanthan gum.[70] Polyacrylamide grafted starch and its applicability as flocculant for water treatment has also been reported.[71] In a study by Zhao *et al.* MW irradiated graft copolymerization of acrylic acid(AA) and acrylamide(AM) on cellulose to prepare super absorbent resin (SAR) was investigated. *N,N*-methylene bis-acrylamide was used as a crosslinking agent, and potassium persulfate/sodium thiosulfate as an initiator. Elimination of the use of nitrogen for protection in the preparation of SAR was the major feature of MW conditions for the reaction. Also, the advantages of a short reaction time, the simple process and the low cost were given.[72]

Electrical conducting and patterned layers of a semiconducting polymer *via* microwave-assisted grafting have also been reported.[73,74] Carter *et al.* grafted poly(9,9'-*n*-dihexyl fluorene) (PDHF) onto flat and nanopatterned crosslinked photopolymer films. It was observed that MW irradiation was introduced as an effective means to drive graft formation and thus allow fabrication of PDHF-functionalized surfaces in as little as 30 min. MW-assisted grafting from these patterned surfaces produced fluorescent features which were imaged by optical microscopy.

2.4 Controlled Radical Polymerization

The controlled radical polymerizations belong to modern synthesis concepts. Controlled/living mechanisms provide simple and robust routes to the synthesis of polymers with predetermined molecular weights, low PDIs, specific functionalities and various architectures. The mechanism proceeds mainly by two processes, namely:

(a) reversible termination;
(b) reversible chain transfer.

The reversible termination procedure includes atom-transfer radical polymerization (ATRP) and nitroxide-mediated radical polymerization (NMP), while the latter comprises reversible addition-fragmentation transfer (RAFT).[75,76] Despite several advantages, the controlled radical polymerization requires a large amount of catalyst, a long reaction time and a high temperature to achieve a higher polymerization rate. In order to overcome these barriers, MW as a source of energy has been widely used. The major advantage of performing reactions under MW conditions is the highly accelerated reactions; this implies that it can enhance the reactivity of the reaction system.[77,78] Zhu and co-workers pioneered MW-assisted controlled radical reactions and their contribution is immense in the field. In one of

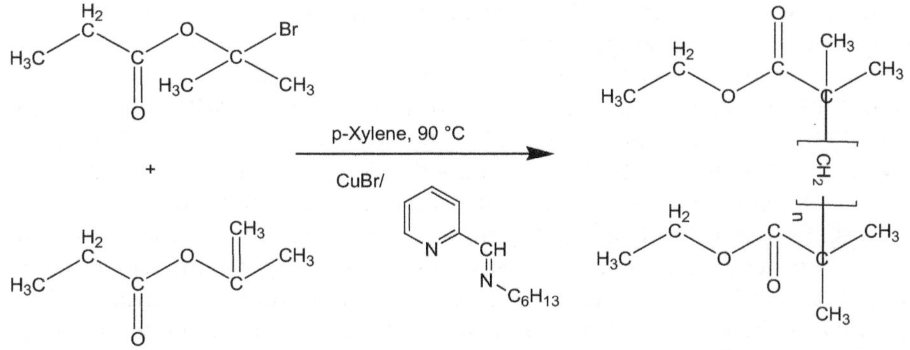

Scheme 2.4 MW-assisted ATRP of MMA.[79]

their initial contributions, they reported the ATRP of MMA with the α,α'-dichloroxylene/CuCl/N,N,N',N'',N''-pentamethyldiethylenetriamine (PMDETA) initiation system under microwave irradiation (Scheme 2.4).[79] Here α,α'-dichloroxylene was used as an initiator and the latter two were used as catalysts. In order to realize the aim of the work *i.e.* to enhance the polymerization rate, decrease the reaction temperature and increase the initiator efficiency in the presence of microwave irradiation, they modified and customized a Samsung domestic microwave oven (Model M9D88). The reaction followed a linear first-order rate, a linear increase of the number-average molecular weight with conversion, and low polydispersity, indicating the controlled nature of the reaction. The authors attributed the 14-fold rate increment to the higher solubility of the copper salt under MW irradiation.

Zhang and Schubert presented a report focusing on the reality of the "microwave effect" in the microwave-assisted ATRP as well as the applicability of the advantage of microwave heating over conventional heating, (for example, the significant increase in reaction rate) to the ATRP system. They also compared their experimental findings with reported literature information.[80] A monomode microwave synthesizer (EmrysTM Liberator, Personal Chemistry Ltd.) was used for ATRP of MMA with CuBr or CuCl/ N-hexyl-2-pyridylmethynimine/ethyl 2-bromoisobutyrate in solution of p-xylene. The authors concluded that, though the reaction successfully proceeded in a controlled manner under microwave irradiation, no significant "microwave effects" were observed. In another work, Zhu and co-workers performed the ATRP of methyl methacrylate in hexane, with the initiator α,α'-dichloroxylene and the catalysts CuCl and N,N,N',N'',N''-pentamethyldiethylenetriamine.[81] The authors highlighted the fact that the use of a thermostat bath does not necessarily reveal the reaction temperatures in the reaction vial, especially if solid ionic particles are involved in the reaction. These ionic particles optimally absorb microwave irradiation because of its electromagnetic character. As a result, the non-contact heating in the microwave reactors might heat the reaction mixture well beyond the temperature of the thermostat bath. One can therefore infer that a direct comparison with findings from reactions performed under conventional heating would be no longer valid.

Wisnoski *et al.* prepared novel resins in a monomodal microwave reactor.[82] They set up a 10 min reaction at 185 °C of TEMPO-methyl resin (TEMPO: 2,2,6,6-tetramethylpiperidine-*N*-oxyl) with various functionalized styrenes or 4-vinyl pyridine and obtained large (>500 mm), spherical resin beads. The authors strongly assumed a controlled radical mechanism (involving the TEMPO radical) in the course of the polymerization which was in fact 150 times faster than with conventional heating. Zelentzova *et al.* observed a re-initiation of the AIBN (2,2'-azoisobutyronitrile) initiated and TEMPO mediated radical polymerization of MMA under microwave irradiation at 70 °C.[83] The radical species were detected by means of electron spin resonance (ESR) spectroscopy. The authors attributed this observation to the decomposition of the intermediately formed non-radical TEMPO adduct with the 2-cyanopropan-2-yl radical.

Demonceau *et al.* reported that the use of MW as a heat source successfully leads to controlled ATRP of MMA catalyzed by $[RuCl_2(p\text{-cymene})(PCy_3)]$ at temperatures lower than 120 °C.[84,85] They demonstrated that, not only temperature, but also the method played a significant role in the reaction course. MW synthesis with simultaneous cooling of the reaction vessel led to a threefold faster rate than the reactions performed in an oil bath. At higher temperatures, however, the polymerizations were no longer controlled and significant differences were found between MW-assisted polymerizations and conventionally heated protocols. This was attributed to a very high concentration of radical species, which resulted in enhanced propagation, termination and mostly disproportionation rates (Scheme 2.5).

Hou *et al.* devised a single-pot microwave irradiated ATRP synthesis of polyacrylonitrile using $FeBr_2$/isophthalic acid as the catalyst and 2-bromopropionitrile as the initiator (Scheme 2.6).[86] The 1:2 ratio of $FeBr_2$ to isophthalic acid not only gave the best control of molecular weight and its distribution but also increased the reaction rate. The polymers obtained were end-functionalized by bromine atoms, and they were used as macroinitiators to continue the chain extension polymerization.

Zhu *et al.* conducted RAFT polymerizations of styrene in bulk under microwave irradiation, with or without AIBN, at 72 and 98 °C, respectively.[87] Structural characterizations using ^{1}H and ^{13}C NMR confirmed the same structure for both polymers obtained under microwave irradiation and conventional heating. The reaction showed living/controlled features, and there was a significant enhancement of the polymerization rates (5.4 and 6.2 times) under microwave irradiation in comparison with conventional heating under the same conditions. Successful chain-extension experimentation further proved the livingness of the RAFT polymerization carried out under microwave. Brown *et al.* set up a system that consisted of AIBN and ethylthiosulfanyl-carbonylpropionic acid ethyl ester at 60 °C, in which only a slight increase in polymerization rate was observed in comparison to conventional heating conditions. In a more recent publication, the same group

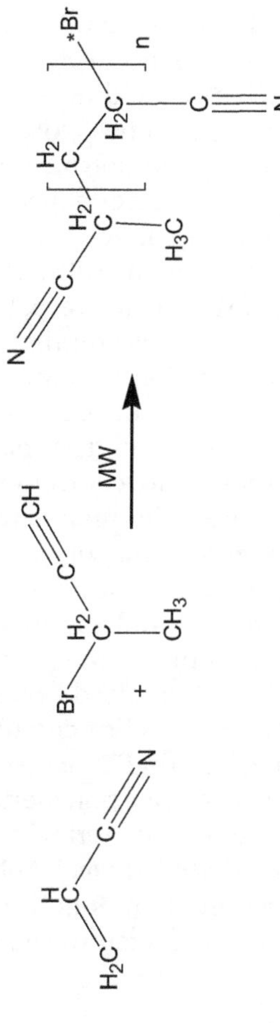

Scheme 2.5 MW-assisted ATRP of MMA catalyzed by [RuCl$_2$(p-cymene)(PCy$_3$)].[84,85]

Scheme 2.6 MW-assisted ATRP of acrylonitrile with FeBr$_2$/isopthalicacid/2-brompropionilrile.[86]

showed that the use of a monomode MW reactor without controlling reaction temperatures led to ultra-fast reaction rates of the RAFT polymerization of MMA, vinyl acetate and styrene. In particular, methyl acrylate and vinyl acetate polymerizations were almost quantitative after 15–20 min, whereby well-defined and well-controlled polymers were obtained.[88,89] Perrier group employed a monomodal microwave reactor for the RAFT-mediated polymerization of MA, MMA and styrene at comparable temperatures, and a uniform power output of between 1 and 3 Watts, the microwave gave better results than the oil bath in terms of polymerization rate for both MA and MMA, showing rate enhancements of up to 152% and 254% for the two monomers, respectively. Furthermore, the control appears unaffected by the rate increase, yielding low polydispersities (PDIs) (as low as 1.05) and molecular-weight values close to predicted values. They also experimentally established that the polymerization rate of styrene, a non polar monomer, does not seem to be enhanced by the use of microwave heating.[89] A comparison between microwave and thermal heating for the RAFT polymerization of MMA with AIBN as initiator and 2-cyano-2-butyldithiobenzoate (CBDB) as RAFT agent, revealed comparable polymerization kinetics indicating the absence of non-thermal microwave effects.[85] The CBDB-mediated RAFT polymerization of MMA was investigated at high temperatures in the absence of a radical initiator, showing a linear increase of the molar masses with conversion, and the polydispersity indices indicated a controlled polymerization. This control over the polymerization was confirmed by the ability to control the molar masses by the concentration of RAFT agent.[88] RAFT polymerization of styrene in the presence of 2-cyanoprop-2-yl 1-dithionaphthalate with and without initiator in a modified domestic MW oven was studied and the living character was retained under the applied reaction conditions.[87] Microwave irradiation was used for the synthesis of side-chain functionalized polystyrene by combination of controlled radical polymerization and click chemistry (Scheme 2.7).[90] The ATRP of acrylonitrile with FeBr$_2$/isophthalic acid/2-bromopropionitrile under MW conditions showed much higher polymerization rate as compared to the rate obtained under conventional refluxing with tetrachloromethane.[86] MW-assisted nitroxide-mediated radical polymerization of acrylamide in aqueous solution was reported by Rigolini *et al.*[91] Facile RAFT precipitation polymerization for the microwave-assisted synthesis of well-defined, double

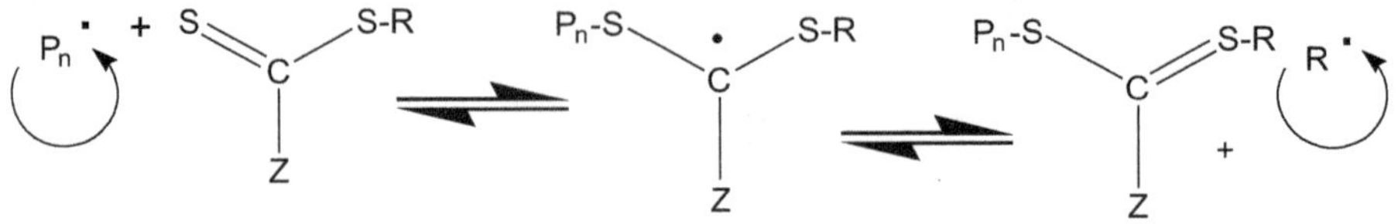

Scheme 2.7 Reaction scheme depicting rapid chain transfer of thiocarbonyl thio moieties between propagating chains for RAFT Xanthate polymerizations.[90]

hydrophilic block copolymers and nanostructured hydrogels was also reported.[92] Copolyacrylates with phenylalanine and anthracene entities were prepared by ATRP and microwave irradiation.[93]

Schubert *et al.* synthesized side-chain functionalized macromolecules by combination of controlled radical polymerization and click chemistry.[94] They obtained alkyne containing polymers with narrow molecular weight distributions (PDI < 1.3) by RAFT polymerization of trimethylsilyl-propargyl methacrylate (TMSPMA). Homopolymers of TMSPMA, as well as random and block copolymers with methyl methacrylate, were synthesized and subsequently deprotected to obtained acetylene-functionalized polymers. These polymers in turn were further functionalized using the copper(I) catalyzed alkyne–azide cycloaddition reaction to obtain well-defined glycopolymers, grafted copolymers and anthracene-containing polymers.

Grassl and co-workers used a combination of microwave irradiation as a heating source and water as a solvent for carrying out a NMP of acrylamide using β-phosphonylated nitroxide.[95] The group also studied the dependence of rate enhancement of polymerization on the mode of irradiation, *i.e.*, either a dynamic (DYN) mode or a pulse (SPS) mode. The former mode corresponded to a dynamic control of the temperature by way of a high initial microwave power, and in this case, no specific microwave effect was observed. On the other hand, in the SPS mode, which is a pulsed power mode, the result showed a strong acceleration of the polymerization process (>50 times) without the loss of the living/controlled polymerization characteristics, which is relevant with a re-initiation of the polyacrylamide macroinitiator even after 100% of conversion. This is depicted in Scheme 2.8.

Nicolay *et al.* and Phan *et al.* were amongst the first to demonstrate NMP in homogeneous aqueous solution below the water boiling point.[96,97] The former used a carboxy-functionalized nitroxide based on the 2,2,5-trimethyl-4-phenyl-3-azahexane-3-oxy (TIPNO) structure for the synthesis of two difunctional alkoxyamines. Due to the presence of the carboxylic acid function, the nitroxide is organo-soluble in its acidic form and water-soluble in its basic form. Polymerizations of styrene and *n*-butyl acrylate mediated with the functional nitroxide had all the expected features of a controlled reaction. The presence of an active chain end was demonstrated by re-initiation of a polystyrene block to form a polystyrene-*b*-poly(*n*-butyl acrylate) block copolymer. Well-defined polymers of sodium styrenesulfonate were successfully synthesized at temperatures below 100 °C.

2.5   Emulsion Polymerization

Emulsion polymerization is a type of radical polymerization that usually starts with an emulsion incorporating water, monomer and surfactant. The most common type of emulsion polymerization is an oil-in-water emulsion, in which droplets of monomer (the oil) are emulsified (with surfactants) in a continuous phase of water. Water-soluble polymers, such as certain

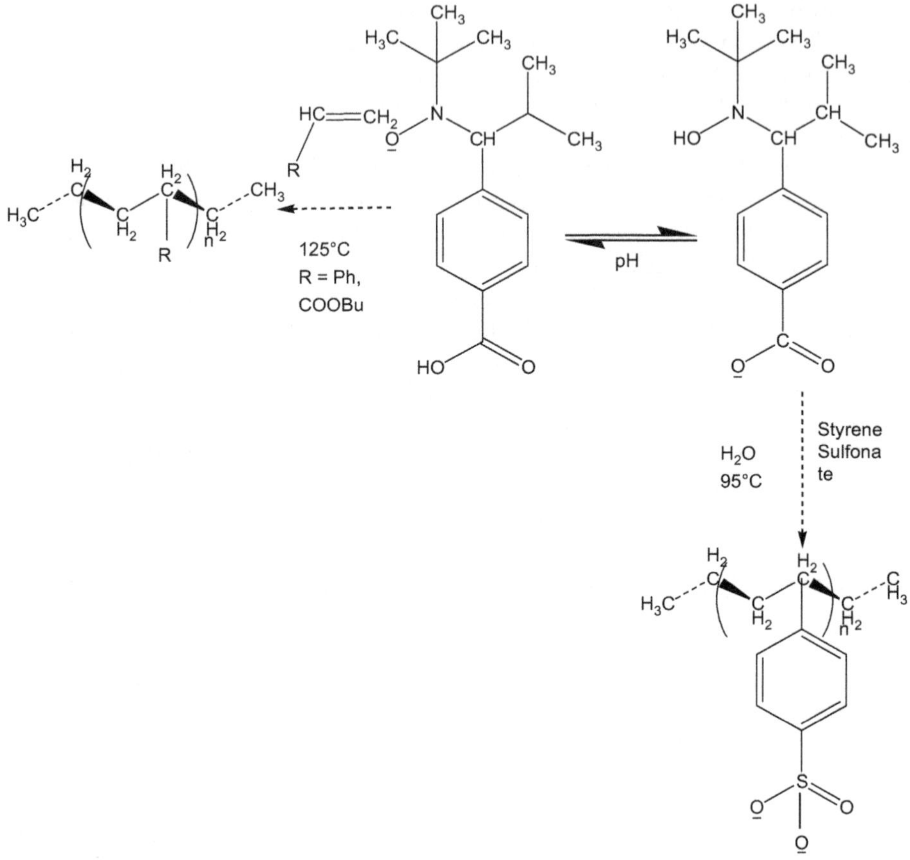

Scheme 2.8   NMP of acrylamide.[95]

polyvinyl alcohols or hydroxyethyl celluloses, can also be used to act as emulsifiers/stabilizers. Polymerization takes place in the latex particles that form spontaneously in the first few minutes of the process, schematically represented in Figure 2.1.

The latex particles are typically 100 nm in size, and are made of many individual polymer chains. The particles are stopped from coagulating with each other because each particle is surrounded by the surfactant ("soap"); the charge on the surfactant repels other particles electrostatically. When water-soluble polymers are used as stabilizers instead of soap, the repulsion between particles arises as these water-soluble polymers form a "hairy layer" around a particle that repels other particles, because pushing particles together would involve compressing these chains.

Emulsion polymerization is used to manufacture several commercially important polymers. Many of these polymers are used as solid materials and must be isolated from the aqueous dispersion after polymerization. In other cases, the dispersion itself is the end product. These emulsions find

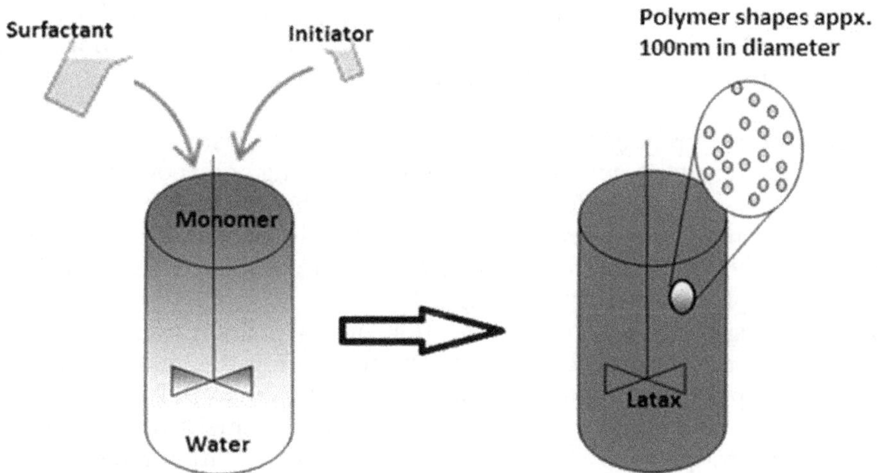

Figure 2.1   Emulsion polymerization: Pictorial presentation.

applications in adhesives, paints, paper coating and textile coatings. They are finding increasing acceptance and are preferred over solvent-based products in these applications as a result of their eco-friendly characteristics due to the absence of VOCs (Volatile Organic Compounds) in them.[75,98,99]

A major challenge in emulsion polymerization is the difficulty of predicting and preparing a batch of latex in terms of particle size distributions and stability because it is not completely clear how each experimental condition affects the emulsion polymerization.[100] The use of microwave radiation as a heating source for emulsion polymerization has been predicted to result in reproducible and narrowly distributed surfactant-free latex particles.[101] Though microwave-assisted emulsion polymerization has been intensively studied, MMA and styrene are investigated most often. Gao and Wu[100] studied the microemulsion and emulsion polymerization of styrene at 70 °C in the presence of sodium dodecyl sulfate (SDS, surfactant) and potassium persulfate (KPS, initiator) under microwave radiation. They characterized the resultant polystyrene latex particles formed at different polymerization stages using laser light scattering. Based on their experimental findings, they demonstrated that stabilization effects of both the surfactant and the ionic end groups are dependent on styrene, SDS and KPS concentration and proposed a simplified model differentiating the process of emulsion polymerization and microemulsion polymerization. The model successfully predicted the effect of monomer concentration on particle size as well as on the number of the resultant latex particles and the effect of diluting the reaction mixture on the resultant particle size. Ngai and Wu reported that emulsifier-free emulsion polymerization of styrene in a water/acetone mixture under microwave irradiation resulted in narrowly distributed stable polystyrene nanoparticles with an average hydrodynamic radius $\langle R_h \rangle$ down to 35 nm when 50 wt% of acetone was added. The process led to a much-desired surfactant-free synthesis where particle size could be controlled by adjusting initial ionic strength.[102] Vivaldo-Lima and co-workers developed a kinetic model for the microwave-assisted initiator free emulsion polymerization of styrene.[103]

Holtze *et al.* investigated microwave-mediated miniemulsion polymerization of styrene with consecutive periods of heating and cooling.[104] In the course of pulsed irradiation of medium hydrophobic azo-initiators under optimized conditions, polymer radicals survive the heating pulse and grow during the cooling period to ultrahigh molecular weights $>10^7$ g/mol. They termed this phenomenon "surviving radical effect" and discovered an unexpectedly high conversion after the first polymerization cycle. This was attributed to the absence of termination step during the cold phase. The temperature pulse comprised a heating phase of less than 10 s duration and a subsequent cooling period. The results proved the merit of this unique combination for polymer synthesis as at high polymerization rates the molecular weight distribution can be tuned over broad ranges. The group further demonstrated, by carrying out surviving radical polymerization in a continuous plug flow tube reactor, that this novel method (Figure 2.2) of pulsed thermal polymerization is suitable as an industrially useful technique. It combines the advantages of an outstanding range of accessible molecular weights, the potential of introducing additives over the course of polymerization, the possibility to carry out a continuous process and an unrivalled space-time yield.[105]

Zhu and co-workers demonstrated the nitroxide-mediated free radical miniemulsion polymerization of styrene with potassium peroxodisulfate as the initiator under microwave conditions at 135 °C.[106] On comparing with experiments under conventional heating, an increased initiator

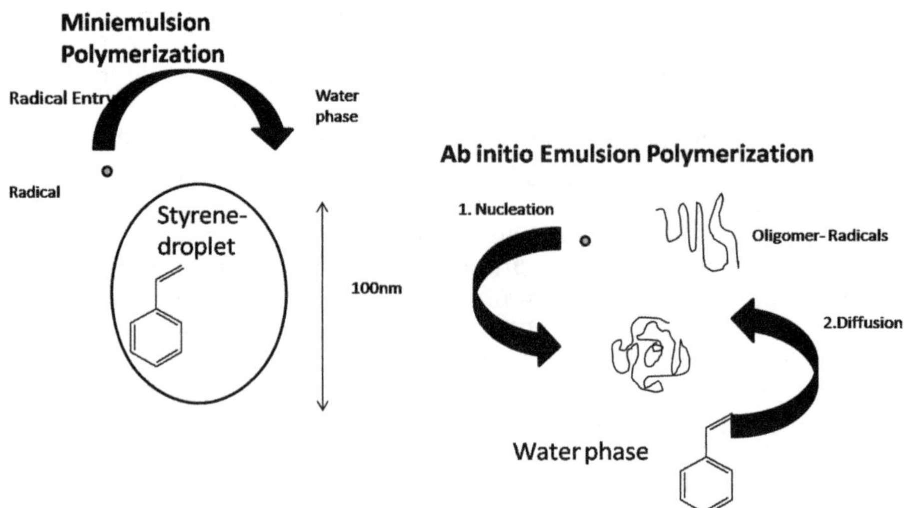

Figure 2.2   Miniemulsion polymerization of styrene.

decomposition was observed under microwave conditions which in turn led to an increased polymerization rate. The detailed kinetic studies revealed that reaction followed a first order kinetics under both the conditions; however, the rate of polymerization was significantly higher under microwave irradiation. The obtained molecular weights showed a linear relationship with respect to the monomer conversion and a good correspondence to the theoretical values.

Bao and Zhang presented a systematic study on surfactant free emulsion polymerization of MMA under microwave irradiation.[107] In order to fully elucidate the relationship between the initiator decomposition and the polymerization rate and to permit a comparison of the microwave results with those obtained under classical conditions, they used a reactor with temperature and power control. They modified a monomodal microwave oven for this purpose. With the aid of transmission electron microscopy they successfully synthesised and characterized PMMA nanoparticles with an average particle size of 45 and 67 nm. Similar results were found by Palacios and co-workers while studying the microwave-assisted (using multimodal reactor) emulsion polymerization of MMA.[108] An acceleration of the polymerization rate by a factor of 137 and polymer samples with a higher number-average molecular weight and lower polydispersity could be obtained under MW conditions.

The preparation of narrow disperse cross-linked PMMA nanoparticles under MW conditions was demonstrated by Hawker and co-workers.[109] Under conditions of super-heated solvent of 25 wt.% acetone/water, a significant reduction of particle size was observed. In addition, a wide range of diameters could be obtained by varying the MW power. PMMA shell on poly(N-isopropylacrylamide) (PNIPAM) particles were prepared by microwave-assisted emulsion polymerization by Zhang and co-workers.[110] The PMMA shell limited the movement of PNIPAM chains, which resulted in a reversible swelling and shrinking. The swelling–shrinking processes of particles were traced by laser light scattering, and the swelling factor of particles was found to be 1.63 with an average radius of 83.4 nm in the swelling state and of 70.8 nm in the shrinking state. Besides this, they measured and analysed the temperature-sensitive property of the particles and concluded that the particles kept the breathing ability over two months, displaying their potential to be an applicable product.

Xu *et al.* successfully carried out emulsifier-free emulsion polymerization of styrene (St) and N-isopropylacrylamide under microwave irradiation, and obtained monodisperse thermoresponsive poly(styrene-*co*-N-isopropylacrylamide) (poly(St-*co*-NIPAM)) particles with diameters in the range 100–130 nm.[111] The morphology, size and size distribution of the poly(St-*co*-NIPAM) particles were characterized by transmission electron microscopy (TEM), scanning electron microscopy (SEM) and photon correlation spectroscopy (PCS), respectively. The results indicated that poly(St-*co*-NIPAM) particles were spherical, and the particles prepared by emulsifier-free emulsion polymerization under microwave irradiation were smaller and more uniform than those obtained with conventional heating. It was reported that the hydrodynamic diameters of poly(St-*co*-NIPAM) particles decreased as the temperature increased from 25 °C to 40 °C, and poly(St-*co*-NIPAM) particles collapsed at 32 °C, which is the lower critical solution temperature of the poly(*N*-isopropylacrylamide). From the SEM micrographs of the assembled poly(St-*co*-NIPAM) particles, it was found that monodisperse poly(St-*co*-NIPAM) particles could assemble to form the two-dimensional order structures. Similar results were observed when monodisperse polystyrene microspheres with diameters of 200–500 nm were prepared by dispersion polymerization with microwave irradiation with poly(*N*-vinylpyrrolidone) as a steric stabilizer and AIBN as a radical initiator in an ethanol/water medium.[112] The microspheres prepared by dispersion polymerization with microwave irradiation were smaller, more uniform and steadier than those obtained with conventional heating.

Shi *et al.* reported an example of water-free two-phase polymerization of a series of poly(*N*-isopropylacrylamide) (PNIPAM)-based hydrogels prepared under microwave irradiation using poly(ethylene oxide)-600 (PEO-600) as a reaction medium and microwave-absorbing agent as well as a pore-forming agent.[113] They presented an interesting study of the swelling behavior of hydrogels using SEM images. They also performed a temperature dependent study in order to ascertain their shelf life for biomedical applications.

Polymers containing rare earth metal ions have been studied extensively for their luminescence applications. Hu *et al.*[114] adopted emulsifier-free emulsion polymerization to synthesize rare earth-containing submicron polymer particles under microwave irradiation. Using potassium persulfate as an initiator and water as a solvent, they studied the polymerization of methyl methacrylate (MMA) in the absence and presence of rare earth ions (europium octanoate (EOA) as source). For particles containing rare earth ions, characterization showed that mole percentage of Eu(III) ions in the surface layer with a thickness of 5 nm which, as estimated from X-ray photo electron spectroscopy (XPS), is larger than the value estimated by inductively coupled plasma atomic emission spectrometer (ICP-AES), indicating that surface enrichment of rare earth ions took place during the polymerization. Albet *et al.* compared thermal and microwave-activated non-aqueous free-radical dispersion polymerization of MMA in *n*-heptane.[114] For controlled polymerization in the microwave field, the group designed a new microwave polymerization reactor to afford uniform heating and to control temperature *via* microwave power variation. At identical MMA concentration and polymerization temperature of 70 °C no special effect of the microwave field on conversions, molecular weights and particle sizes of PMMA was detected with respect to conventional thermal free radical dispersion polymerization.

Luo and co-workers used emulsion polymerization in combination with sol–gel technique to synthesize TiO$_2$/polystyrene core–shell nanospheres with a nano-scale TiO$_2$ core and a smooth and well-defined polystyrene shell.[115] The core-shell morphology was established by TEM images, the structural characterization was done with the help of FT-IR, while the diameter and distribution of the nanospheres were measured by dynamic light scattering. It was inferred that the diameter and its distribution of the TiO_2/polystyrene core–shell nanospheres could be regulated by the concentration of styrene monomer in the emulsion solution. The Huang group used microwave-assisted emulsion polymerization to prepare magnetic Fe_3O_4/poly(styrene-*co*-acrylamide) composite nanoparticles.[116] The reaction was mediated by a ferrofluid coated with oleic acid (OA) and sodium dodecyl sulfate (SDS). The authors studied the influence of monomer concentration and the ferrofluid content in detail. The synthesized magnetic nanoparticles were characterized by FT-IR and X-ray diffraction. The saturation of the magnetic nanoparticles was characterized by a vibrating-sample magnetometer (VSM). TEM and XRD revealed that the Fe_3O_4 nanoparticles were incorporated into the shells of poly(Styrene-*co*-acrylamide). Thermal studies showed that the maximum value observed for weight percent of iron oxide was 45.9%. The magnetic nanoparticles could easily be separated in a magnetic field.

Microwave-assisted radical polymerization has been extensively studied in the last decade. Almost all the studies focused on aspects of speed and reproducibility. The data consistently pointed out to MW as a novel and powerful tool for microwave-assisted synthesis. Scalability of reactions at industrial level is the need for future research in this area.

References

1. Odian and George, *Principles of Polymerization*, Wiley-Interscience, New York, 4th edn, 2004.
2. J. M. G. Cowie and V. Arrighi, *Polymers: Chemistry and Physics of Modern Materials*, CRC Press, Scotland, 3rd edn, 2008.
3. A. Loupy, A. Petit, J. Hamelin, F. Texier-Boullet, P. Jacquault and D. Mathe, New solvent free organic synthesis using focused microwaves, *Synthesis*, 1998, **9**, 1213.
4. G. Moad and D. H. Solomon, *The Chemistry of Free Radical Polymerization*, Pergamon, Oxford, England, 1995.
5. A. Gourdenne, A.-H. Maassarant and P. Monchaux, *Polym. Prepr. (Am. Chem. Soc. Div. Polym. Chem.)*, 1979, **20**, 471.
6. M. Teffal and A. Gourdenne, *Eur. Polym. J.*, 1983, **19**, 543.
7. G. Madras and V. Karmore, *Polym. Int.*, 2001, **50**, 1324.
8. J. Jovanovic and B. Adnadjevic, *J. Appl. Polym. Sci.*, 2007, **104**, 1775.
9. J. Jacob, L. H. L. Chia and F. Y. C. Boey, *J. Appl. Polym. Sci.*, 1997, **63**, 787.
10. J. Jacob, F. Y. C. Boey and L. H. L. Chia, *Ceramic Trans.*, 1997, **80**, 417.
11. S. Sinnwell and H. Ritter, *Macromol. Rapid Commun.*, 2005, **26**, 160.
12. C. Goretzki, A. Krlej, C. Steffens and H. Ritter, *Macromol. Rapid Commun.*, 2004, **25**, 513.
13. M. Iannelli and H. Ritter, *Macromol. Chem. Phys.*, 2005, **206**, 349.
14. M. Iannelli, V. Alupei and H. Ritter, *Tetrahedron*, 2005, **61**, 1509.

15. S. Schmitz and H. Ritter, *Macromol. Rapid Commun.*, 2007, **28**, 2080.

16. V. Singh, A. Tiwari, P. Kumari and A. K. Sharma, *J. Appl. Polym. Sci.*, 2007, **104**, 3702.

17. F. Fischer, R. Tabib and R. Freitag, *Eur. Polym. J.*, 2005, **41**, 403.

18. J. Lu, S.-j. Ji, J. Wu, F. Guo, H. Zhou and X. Zu, *J. Appl. Polym. Sci.*, 2004, **91**, 1519.

19. S. P. Sitaram and J. O. Stoffer, *Polym. Mater. Sci. Eng*, 1993, **69**, 382.

20. J. O. Stoffer and S. P. Sitaram, *Polym. Mater. Sci. Eng.*, 1994, **71**, 55.

21. L. H. L. Chia, J. Jacob and F. Y. C. Boey, *J. Polym. Sci., Part A: Polym. Chem.*, 1996, **34**, 2087.

22. M. A. Dikusar, I. V. Kubrakova, A. A. Chinarev and N. V. Bovin, *Russ. J. Bioorg. Chem.*, 2001, **27**, 408.

23. M. S. Cortizo, *J. Appl. Polym. Sci.*, 2007, **103**, 3785.

24. M. S. Cortizo, S. Laurella and J. L. Alessandrini, *Radiat. Phys. Chem.*, 2007, **76**, 1140.

25. Y. Chen, J. Wang, D. Zhang, R. Cai, H. Yu, C. Su and Z.-E. Huang, *Polymer*, 2000, **41**, 7877.

26. A. F. Porto, B. L. Sadicoff, M. C. V. Amorim and M. C. S. Mattos, *Polym. Test.*, 2002, **21**, 145.

27. H. F. Naguib, G. R. Saad and M. Z. Elsabeé, *Polym. Int.*, 2003, **52**, 1217.

28. T. Biswal, R. Samal and P. K. Sahoo, *J. Appl. Polym. Sci.*, 2010, **117**, 1837.

29. H. Stange, M. Ishaque, N. Niessner, M. Pepers and A. Greiner, *Macromol. Rapid Commun.*, 2006, **27**, 156.

30. H. Stange and A. Greiner, *Macromol. Rapid Commun.*, 2007, **28**, 504.

31. C. M. Fellows, *Cent. Eur. J. Chem.*, 2005, **3**, 40.

32. S. Agarwal, M. Becker and F. Tewes, *Polym. Int.*, 2005, **54**, 1620.

33. J. M. Lu, J. F. Wu, L. H. Wang and S. H. Ya, *J. Appl. Polym. Sci.*, 2005, **97**, 2072.

34. J. Palacios, J. Sierra and P. Rodriguez, *Polym. Prepr. (Am. Chem. Soc. Div. Polym. Chem.)*, 1991, **32**, 244.

35. J. Palacios, J. Sierra and P. Rodriguez, *New Polym. Mater.*, 1992, **3**, 273.

36. J. Lu, X. Zhu, J. Zhu and J. Yu, *J. Appl. Polym. Sci.*, 1997, **66**, 129.

37. J. Lu, X. Zhu, S. Ji, J. Zhu and Z. Chen, *J. Appl. Polym. Sci.*, 1998, **67**, 1563.

38. J. Lu, Q. Jiang, X. Zhu and F. Wang, *J. Appl. Polym. Sci.*, 2001, **79**, 312.

39. J. Lu, S. Ji, J. Wu and X. Zhu, *Chem. J. Internet*, 2001, **3**, 41.

40. J. Lu, S. Ji, J. Wu, F. Guo, H. Zhuo, W. Dai and X. Zhu, *J. Macromol. Sci. Pure. Appl. Chem.*, 2002, **A39**, 351.

41. J. Lu, S. Ji, J. Wu, F. Guo, H. Zhuo and X. Zhu, *J. Appl. Polym. Sci.*, 2004, **91**, 1519.

42. R. Gref, J. Rodrigues and P. Couvrceur, *Macromolecules*, 2002, **35**, 9861.

43. Y. Ohya, S. Maruhashi, T. Hirano and T. Ouchi, in *Biomedical Polymers and Polymer Therapeutics*, ed. E. Chiellini, J. Sunamoto, C. Migliaresi, R. M. Ottenbrite and D. Cohn, Kluwer Academic/Plenum Publishers, New York, 2001, p. 139.

44. M. Tizzotti, A. Charlot, E. Fleury, M. Stenzel and L. Bernard, *Macromol. Rapid Commun.*, 2010, **31**, 1751.

45. http://www.cmu.edu/maty/index.html.

46. V. Singh, A. Tiwari, D. N. Tripathi and R. Sanghi, *Carbohydr. Polym.*, 2004, **58**, 1.

47. Z. Wan, Z. Xiong, H. Ren, Y. Huang, H. Liu, H. Xiong, Y. Wu and Han, *J. Carbohydr. Polym.*, 2011, **83**, 264.

48. B. S. Kaith, R. Jindal, A. K. Jana and M. Maiti, *Iran. Polym. J.*, 2009, **18**, 789.

49. G. Sen, R. V. Kumar, S. Ghosh and S. Pal, *Carbohydr. Polym.*, 2009, **77**, 822.

50. V. Singh and A. Tiwari, *Int. J. Biol. Macromol.*, 2009, **44**, 186.

51. V. Singh, A. Tiwari, P. Kumari and S. Tiwari, *Carbohydr. Res.*, 2006, **341**, 2270.

52. V. Singh and A. Tiwari, *Carbohydr. Res.*, 2008, **343**, 151.

53. D. Stuerga, K. Gonon and M. Lallemant, *Tetrahedron*, 1993, **49**, 6229.

54. V. Singh, A. Tiwari and R. Sanghi, *Prog. Polym. Sci.*, 2012, **37**, 340.

55. V. Singh, D. N. Tripathi, A. Tiwari and R. Sanghi, *Carbohydr. Polym.*, 2006, **65**, 35.

56. V. Singh, A. Tiwari, D. N. Tripathi and R. Sanghi, *Polymer*, 2006, **47**, 254.

57. V. Singh, D. N. Tripathi, A. Tiwari and R. Sanghi, *J. Appl. Polym. Sci.*, 2005, **95**, 820.

58. V. Singh, A. Tiwari, D. N. Tripathi and R. Sanghi, *J. Appl. Polym. Sci.*, 2004, **92**, 1569.

59. V. Singh and D. N. Tripathi, *J. Appl. Polym. Sci.*, 2006, **101**, 2384.

60. V. Singh, A. Tiwari, S. Pandey and S. K. Singh, *Starch-Stärke*, 2006, **58**, 536.

61. K. Prasad, R. Meena and A. K. Siddhanta, *J. Appl. Polym. Sci.*, 2006, **101**, 161.

62. R. Meena, K. Prasad, G. Mehta and A. K. Siddhanta, *J. Appl. Polym. Sci.*, 2006, **102**, 5144.

63. Z. Tong, W. Peng, Z. Zhiqian and Z. Baoxiu, *J. Appl. Polym. Sci.*, 2005, **95**, 264.

64. K. Xu, W. D. Zhang, Y. M. Yue and P. X. Wang, *J. Appl. Polym. Sci.*, 2005, **98**, 1050.

65. J. M. Pawluczyk, R. T. McClain, C. Denicola, J. J. Mulhearn, D. J. Rudd and C. W. Lindsley, *Tetrahedron Lett.*, 2007, **48**, 1497.

66. J. Zhang, S. Zhang, K. Yuan and Y. Wang, *J. Macromol. Sci., Pure Appl. Chem.*, 2007, **44**, 881.

67. V. Singh, P. L. Kumari, A. Tiwari and A. K. Sharma, *Polym. Adv. Technol.*, 2007, **18**, 379.

68. A. Mishra, J. H. Clark, A. Vij and S. Daswal, *Polym. Adv. Technol.*, 2007, **18**, 1.

69. B. S. Kaith, R. Jindal, K. Janaasim and M. Malti, *Int. J. Polym. Anal. Charact.*, 2009, **14**, 364.

70. A. Kumar, K. Singh and M. Ahuja, *Carbohydr. Polym.*, 2009, **76**, 261.

71. S. Mishra, A. Mukul, G. Sen and U. Jha, *Int. J. Biol. Macromol.*, 2011, **48**, 106.
72. J. Shu, X. J. Li and D. B. Zhao, *Adv. Mater. Res.*, 2010, **148**, 799.
73. A. Tiwari and V. Singh, *Carbohydr. Polym.*, 2008, **74**, 427.
74. K. R. Carter, S. B. Jhaveri and I. W. Moran, *Small*, 2008, **4**, 1176.
75. S. Sinnwell and H. Ritter, *Aust. J. Chem.*, 2007, **60**, 729.
76. S. Delfosse, Y. Borguet, L. Delaude and A. Demonceau, *Macromol. Rapid Commun.*, 2007, **28**, 492.
77. S. Caddick, *Tetrahedron*, 1995, **51**, 10403.
78. Z. P. Cheng, X. L. Zhu, G. J. Chen, W. J. Xu and J. M. Lu, *J. Polym. Sci., Part A: Polym. Chem.*, 2002, **40**, 3823.
79. X. Li, X. Zhu, Z. Cheng, W. Xu and G. Chen, *J. Appl. Polym. Sci.*, 2004, **92**, 2189.
80. H. Zhang and U. S. Schubert, *Macromol. Rapid Commun.*, 2004, **25**, 1225.
81. X. Li, X. Zhu, Z. Cheng, W. Xu and G. Chen, *J. Appl. Polym. Sci.*, 2004, **92**, 2189.
82. D. D. Wisnoski, W. H. Leister, K. A. Strauss, Z. Zhao and C. W. Lindsley, *Tetrahedron Lett.*, 2003, **44**, 4321.
83. N. V. Zelentzova, Y. D. Semchikov, N. A. Kopylova, M. V. Kuznetsov, S. V. Zelentsov and A. N. Konev, *J. Polym. Sci., Part B: Polym. Phys.*, 2003, **45**, 65.
84. S. Delfosse, H. Wei, A. Demonceau and A. F. Noels, *Polym. Prepr. (Am. Chem. Soc., Div. Polym. Chem.)*, 2005, **46**, 295.
85. S. Delfosse, Y. Borguet, L. Delaude and A. Demonceau, *Macromol. Rapid Commun.*, 2007, **28**, 492.
86. C. Hou, Z. Guo, J. Liu, L. Ying and D. Geng, *J. Appl. Polym. Sci.*, 2007, **104**, 1382.
87. J. Zhu, X. Zhu, Z. Zhang and Z. Cheng, *J. Polym. Sci., Part A: Polym. Chem.*, 2006, **44**, 6810.
88. S. L. Brown, C. Rayner and M. S. Perrier, *Macromol. Rapid Commun.*, 2007, **28**, 478.
89. S. L. Brown, C. Rayner, M. S. Graham, A. Cooper, S. Rannard and S. Perrier, *Chem. Commun.*, 2007, **21**, 2145.
90. C. R. Krieg, R. Becer, R. Hoogenboom and U. S. Schubert, *Macromol. Symp.*, 2009, **275**, 73.
91. J. Rigolini, B. Grassl, S. Reynaud and L. Billon, *J. Polym. Sci., Part A: Polym.Chem.*, 2010, **48**, 5775.
92. Z. An, Q. Shi, W. Tang, C. Tsung, C. J. Hawker and G. D. Stucky, *J. Am. Chem. Soc.*, 2007, **127**, 14493.
93. E. C. Buruiana, M. Murariu and T. Buruiana, *J. Lumin.*, 2010, **130**, 1794.
94. K. Babiuch, C. R. Becer, M. Gottschaldt, J. T. Delaney, J. R. Weisser, B. Beer, R. Wyrwa, M. Schnabelrauch and U. S. Schubert, *Macromol. Biosci.*, 2011, **11**, 535.
95. J. Rigolini, B. Grassl, L. Billon, S. Reynaud and O. F. X. Donard, *J. Polym. Sci., Part A: Polym. Chem.*, 2009, **147**, 6919.

96. R. Nicolay, L. Marx, P. Hemery and K. Matyjaszewski, *Macromolecules*, 2007, **40**, 6067.

97. T. N. T. Phan and D. Bertin, *Macromolecules*, 2008, **41**, 1886.

98. Terminology of polymers and polymerization processes in dispersed systems (IUPAC Recommendations 2011), *Pure Appl. Chem.*, 2011, **83**, 2229.

99. C. Ebner, T. Bodner, F. Stelzer and F. Wiesbrock, *Macromol. Rapid Commun.*, 2011, **32**, 254.

100. J. Gao and C. Wu, *Langmuir*, 2008, **21**, 782.

101. W. M. Zhang, J. Gao and C. Wu, *Macromolecules*, 1997, **30**, 6388.

102. T. Ngai and C. Wu, *Langmuir*, 2005, **21**, 8520.

103. M. A. Aldana-Garcia, J. Palacios and E. Vivaldo-Lima, *J. Macromol. Sci., Part A: Pure Appl. Chem.*, 2005, **42**, 1207.

104. C. Holtze, M. Antonietti and K. Tauer, *Macromolecules*, 2006, **39**, 5720.

105. C. Holtze and K. Tauer, *Macromol. Rapid Commun.*, 2007, **28**, 428.

106. J. Li, X. Zhu, J. Zhu and Z. Cheng, *Radiat. Phys. Chem.*, 2007, **76**, 23.

107. J. Bao and A. Zhang, *J. Appl. Polym. Sci.*, 2004, **93**, 2815.

108. J. Sierra, J. Palacios and E. Vivaldo-Lima, *J. Macromol. Sci., Part A: Pure Appl. Chem.*, 2006, **43**, 589.

109. Z. An, W. Tang, C. J. Hawker and G. D. Stucky, *J. Am. Chem. Soc.*, 2006, **128**, 15054.

110. H. Zhao, H. W. Chen, Z. C. Li, W. Su and Q. J. Zhang, *Eur. Polym. J.*, 2006, **42**, 2192.

111. C. F. Yi, Z. W. Deng and Z. S. Xu, *Colloid Polym. Sci.*, 2005, **283**, 1259.

112. Z. S. Xu, Z. W. Deng, X. X. Hu, L. Li and C. F. Yi, *J. Polym. Sci., Part A: Polym. Chem.*, 2005, **43**, 2368.

113. S. Shi and L. Liu, *J. Appl. Polym. Sci.*, 2006, **102**, 4177.

114. J. Hu, H. Zhao, Q. Zhang and W. He, *J. Appl. Polym. Sci.*, 2003, **89**, 1124.

115. H. L. Luo, J. Sheng and Y. Z. Wan, *Mater. Lett.*, 2008, **62**, 37.

116. J. Huang, H. Penb, Z. Xu and C. Yi, *React. Funct. Polym*, 2008, **68**, 332.

CHAPTER 3

Step-growth Polymerization

3.1 Introduction

Step-growth polymerization is a type of polymerization in which bifunctional or multifunctional monomers react successively to form dimers, then trimers, longer oligomers and eventually long chain polymers. The mechanism boasts of synthesis of several naturally occurring and some synthetic polymers like polyesters, polyamides, polyurethanes *etc.* If there are two reactive sites on a monomer, branched polymers are produced. Step-growth polymerization is often compared and contrasted with chain-growth polymerization, as both involve step by step addition of monomer units (Table 3.1).

The history of step-growth polymerization is as old as the history of synthetic polymers itself. The first truly synthetic polymeric material, Bakelite, was prepared by Leo Baekeland in 1907, through a typical step-growth polymerization reaction of phenol and formaldehyde.[1] As step-growth polymerization reactions proceed between two different functional groups, several types of chemical reactions can be used to synthesize polymeric materials by this method. These reactions are based on the coupling of two multifunctional, mostly bifunctional, monomers. The resulting coupled product will naturally contain the functional groups and thus react in the same manner as the monomer, in turn eventually leading to polymeric materials. The most studied step-growth polymerization methods are better known as polycondensations due to the release of water during the coupling reactions.[2]

Step-growth polymerization is amongst the first polymerizations to be studied under MW irradiation. The mechanism became hugely popular on account of the highly reduced reaction timings. The popularity of MW irradiated step-growth polymerization reactions has been very well documented in several review articles and books.[2–7] The development of efficient and mild methods for the synthesis of condensation polymers is a

RSC Green Chemistry No. 35
Microwave-Assisted Polymerization
By Anuradha Mishra, Tanvi Vats and James H. Clark
© Anuradha Mishra, Tanvi Vats and James H. Clark 2016
Published by the Royal Society of Chemistry, www.rsc.org

Table 3.1 Differences between chain-growth and step-growth polymerization.

Step-growth polymerization	Chain-growth polymerization
(a) Polymerization involves growth throughout matrix	The process involves growth by addition of monomer only at one end or both ends of chain
(b) Rapid loss of monomer is experienced early in the reaction	Some amount of monomer concentration remains even at long reaction times
(c) The reaction proceeds *via* similar repetitive steps throughout the reaction	The reaction involves distinct initiation, propagation, termination and chain transfer steps
(d) High extents of reaction are required to obtain high chain length	Molar mass of backbone chain increases rapidly at early stage which remains approximately the same throughout
(e) There is no termination step and the ends remain active	The reaction involves proper termination step and chains are inactive after that
(f) The reaction usually do not require initiator	Initiator is more often than not mandatory

significant aspect of polymer chemistry. In the field of MW-assisted step-growth polymerizations, esterification, amidation, imidation and metal-catalyzed cross-coupling reactions have been studied in recent years.[8] Some important step-growth polymerization reactions are discussed in the following sections.

3.2 Synthesis of Poly(amide)s

Amide linkages can be labeled as the most abundant linkages in nature, courtesy of their involvement in peptides, proteins and enzymes. The evolution of life is due to the formation of an assembly of extrinsically directed amino acid sequences from the assembly of non-directed amino acid sequences. In order to decipher the concepts of evolution in 1990, Yanagawa *et al.* subjected amino acid amides to repeated hydration–dehydration cycles in a domestic microwave oven.[9] They composed a stock mixture of equimolar amounts of glycinamide, alaninamide, valinamide and aspartic acid α-amide as model substances for molecules formed at an early stage of the chemical evolution. After continued solvation in aqueous solution and subsequent evaporation of the water in a microwave oven, formation of polypeptides with molecular weights of up to 4000 daltons was observed. The authors mentioned that the molecular weight of the resulting polypeptide was independent of the number of hydration–dehydration cycles and a saturating yield of 10% was obtained after ten cycles. It was observed that the MW irradiations did not affect the maximum molecular weight attainable, but improved the yield of the polypeptides by a factor of 100. The authors attributed this success to the shortness of the heating periods, which kept undesired side reactions, such as the hydrolysis of the terminal active amide groups, to a minimum.

Imai *et al.* described the MW approach to aliphatic polymers with amide linkages[10] from polycondensation of amino acids and nylon salts. They

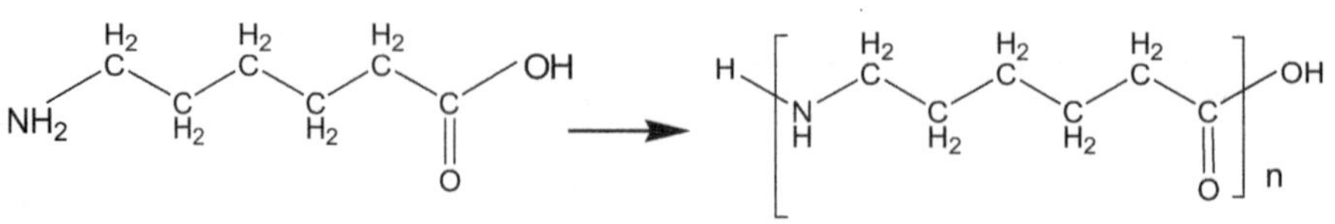

Scheme 3.1 Synthesis nylon6 from ε-amino caproic acid ($n = 5$).[11,12]

derived artificial polyamides from amino acids as well as nylon salts. With the aid of a domestic microwave oven, Imai *et al.* polymerized amino acids of the composition $H_2N(CH_2)_xCOOH$ ($x = 5$, 6, 10, 11, 12) (Scheme 3.1).[11,12]

It is worthy to note that, as the monomers used could not absorb MW irradiation, the authors carried out polymerization reactions in a solution in which the solvent additionally acts as absorber. The solvents with high boiling points and high dielectric constants, like *m*-cresol, *o*-chlorophenol, ethylene glycol, sulfolane and *N*-cyclohexyl-2-pyrrolidone, were commonly used. The reaction process involved the syntheses of the polyamides in open reaction vials so that the solvents can evaporate; within five minutes, polymers with large inherent viscosities of around 0.5 dLg^{-1} were obtained. The authors drew an interesting contrast between the type and amount of solvent. The type of solvent used had no influence on the inherent viscosity (and therefore, neither the molecular weight) of the polymers. The amount of solvent, on the other hand, played a role. The use of a smaller amount of solvent resulted in a higher final temperature in a shorter reaction time, affording polymers with higher inherent viscosities.

Polymerizations of aliphatic diamines and dicarboxylic acids for the preparation of polymers of the composition $[-NH(CH_2)_xNHCO(CH_2)_yCO-]n$ (with the combinations $x/y = 6/4$, 6/6, 6/8, 6/10, 8/4, 12/4, 12/6, 12/8, 12/10)[11–13] as well as the polymerizations of aromatic diamines (like 4,4′-methylenedianiline, 4,4′-oxydianiline, 1,3- and 1,4-phenylenediamine) and dicarboxylic acids (like isophthalic acid and terephthalic acid) to yield polymers of the structure $[-NH-Ar-NHCO-Ar'-CO-]n$ after a Yamazaki phosphorylation reaction (Scheme 3.2a and b) have also been studied.[14]

The authors customized a domestic microwave reactor for polymerizations involving aromatic monomers by additionally using a Teflon insulated thermocouple. These polymerization reactions were also carried out in high-boiling solvents with high dielectric constants. The products were obtained after short reaction times (less than one minute) in high yields (85–96%), and the polyamides exhibited medium to high inherent viscosities (up to 0.86 dLg^{-1}).[13] On investigating the difference between continuous and periodic microwave irradiation, it was found that the polymerizations carried out under periodic microwave irradiation allowed for easier temperature control and gave polymers with a higher inherent viscosity as compared with continuous heating, for one series of aliphatic diamines and dicarboxylic acids ($x = 6$, 8, 12; $y = 4$). These findings played a key role in the development of the modern monomodal microwave systems available today in dynamic field tuning.

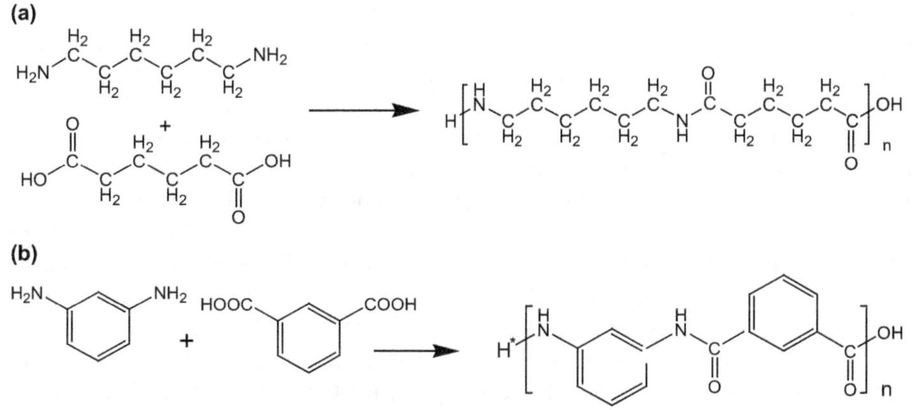

(a)

(b)

Scheme 3.2 (a) MW-assisted polymerization of nylon 6,6 from adipic acid and hexamethyl diamine; (b) Nomex from isopthalic acid and 1,3-methylene diamine.[14]

Pourjavadi *et al.*[15] investigated the microwave-assisted polymerization of linear aliphatic dicarboxylic acids with aromatic diamines, such as *p*-phenylenediamine or 2,5-bis(4-aminophenyl)-3,4-diphenylthiophene by the Yamazaki phosphorylation reaction. They carried out polymerization in the presence of *N*-methylpyrrolidinone (NMP) as a solvent and in a vessel of polyethylene screw-capped cylinder. The polymers obtained after short reaction times (30 or 40 s) with medium to high yields (60–100%) had intrinsic viscosities in the range of 0.10–0.80 dL/g.[16]

Faghihi *et al.* synthesized eight new polyamides containing azo moieties and hydantoin groups under MW irradiation using a domestic microwave oven.[16] They applied polycondensation reactions to 4,4′-azodibenzoyl chloride [4,4′-azobenzenedicarboxylic acid] with eight different derivatives of 5,5-disubstituted hydantoin using *o*-creso as a polar organic medium. Polyamides with inherent viscosity between 0.35 and 0.60 dL/g were obtained in high yields within 7–12 min of reaction time. Elemental analysis, viscosity measurements, thermal gravimetric analysis (TGA and DTG), solubility test and FT-IR spectroscopy were used as tools to characterize the resulting samples. The polyamides exhibited solubility in polar solvents such as *N*,*N*-dimethylacetamide, *N*,*N*-dimethylformamide, dimethylsulfoxide, tetrahydrofurane and *N*-methyl-2-pyrrolidone at room temperature.

It was also reported that the molecular weight of polymers obtained under microwave heating was much higher than the conventional conditions such as interfacial polymerization or the Higashi process. The polymers obtained possessed inherent viscosities between 0.22 and 0.73 dL/g, corresponding to MWs up to 140 000 g/mol, while with interfacial polymerization or the Higashi method, polymers having inherent viscosities in the range of 0.04–0.36 dL/g were formed.[17,18]

Mallakpour and co-workers were pioneers in the study of photoactive polyamides (PAs). Such polyamides containing acetoxynaphthalamide side chain with inherent viscosities of 0.27–0.56 dl g^{-1} were prepared by the direct polycondensation reaction of the 5-(3-acetoxynaphthoylamino)isophthalic acid with various commercially available diamines by means of triphenyl phosphite (TPP) and pyridine (Py) in the presence of calcium chloride and *N*-methyl-2-pyrrolidone (NMP) under both MW irradiation and conventional heating conditions.[19] From Thermo-gravimetric analysis (TGA), it was inferred that polymers were thermally stable. The results showed 10% weight loss temperatures in excess of 320 and 378 °C, and char yields at 600 °C in nitrogen higher than 60%. The absorption study of these macromolecules exhibited maximum UV-Vis absorption at 265 and 300 nm in a DMF solution. Their photoluminescence in the DMF solution demonstrated fluorescence emission maxima around 361 and 427 nm. The same group reported formation of optically active polyamides from reaction of an optically active isosorbide-derived diacyl chloride with two aromatic diamines, diphenylamino-isosorbide (DAI) and 4,4′-diaminodiphenylsulphone (DDS) in NMP under MW irradiation and studied interfacial polymerization. The 6 min of reaction time gave 52% and 70% yields with inherent viscosities of 0.11 and 0.92 dLg^{-1} for diamines of DAI and DDS, respectively, under MW irradiations.[18] They also reported a three-step synthesis of 5-(3-Methyl-2-phthalimidylpentanoylamino)isophthalic acid as a novel aromatic diacid monomer. The first step comprised reaction of phthalic anhydride with L-isoleucine in acetic acid solution. In the second step the resulting imide acid was treated with excess thionyl chloride to give aliphatic acid chloride. The last step was the reaction of this acid chloride with 5-aminoisophthalic acid to yield a bulky chiral aromatic diacid monomer. The direct polycondensation reactions of this diacid with several aromatic and aliphatic diisocyanates such as 4,4′-methylenebis(phenyl isocyanate), toluylene-2,4-diisocyanate, isophorone diisocyanate and hexamethylene diisocyanate were carried out under both MW and classical heating. The comparative account of the polymerization reactions showed that the MW conditions gave faster results and produced a series of novel optically active polyamides (PA)s containing pendent phthalimide group, with good yields and moderate inherent viscosities of 0.17–0.60 dLg^{-1}.[19] The reaction is depicted in Scheme 3.3.

Efficient and rapid synthesis of optically active polyamides in the presence of tetrabutylammonium bromide as ionic liquids under microwave irradiation has also been reported by the same group.[20] They also reported synthesis of polyamide containing flexible l-leucine amino acid.[21] The resulting novel optically active polyamides have inherent viscosities in the range of 0.25–0.63 dl/g. They show good thermal stability and are soluble in amide-type solvents. Microwave irradiation has been established as a versatile tool for increasing reaction rates and yields in synthesis of optically active polyamides.

Taking the MW-assisted polyamide synthesis a green step further introduced MW-assisted synthesis of soluble new optically active polyamides derived from 5-(4-methyl-2-phthalimidylpentanoylamino)isophthalic acid

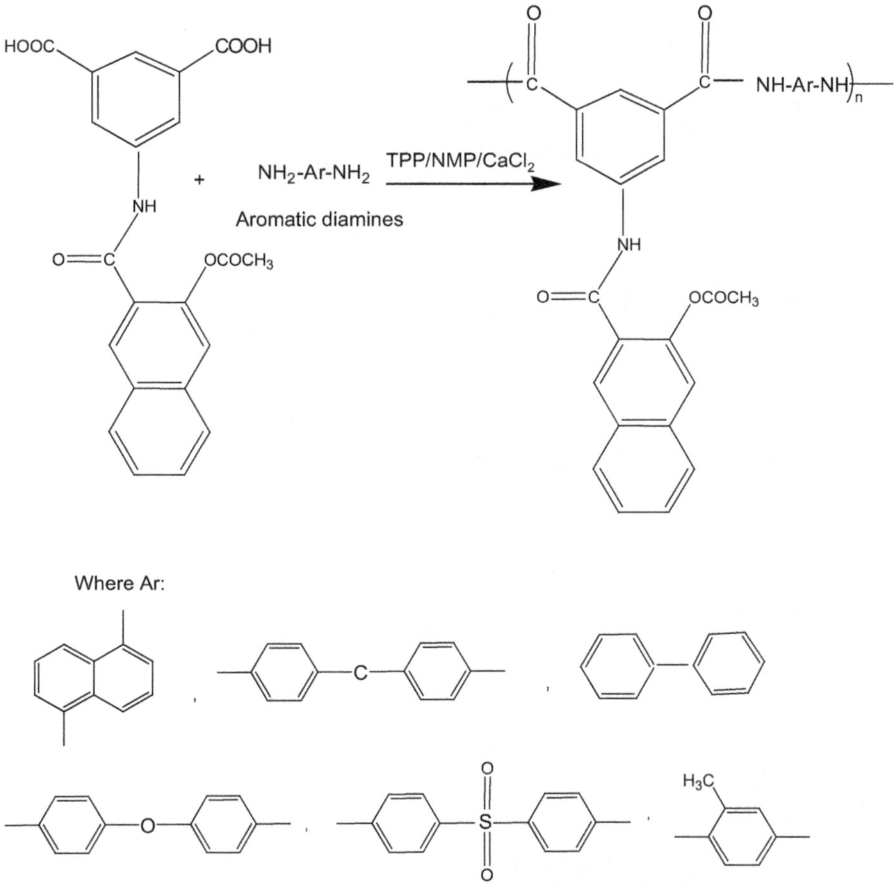

Scheme 3.3 Polyamidation reactions of 5-(3-acetoxynapthoylamino) isopthalic di-acids with aromatic diamines.[19]

different diisocyanates (Scheme 3.4) in molten Ionic liquids (IL)s. Ttetrabutylammonium bromide was used as a molten IL in the presence of different catalysts under microwave irradiation as well as conventional heating. The resulting polyamides (PA)s were characterized by FTIR and ^{1}H NMR spectroscopy, inherent viscosity measurements, thermal and elemental analysis. The obtained PAs showed high yields and moderate inherent viscosities in the range of 0.32–0.57 dL g^{-1}. The PAs were soluble in aprotic polar solvents. The authors highlighted that since toxic and volatile solvent such as NMP were eliminated, this process was safe and green. The combination of IL and microwave irradiation led to large reductions in reaction times, very high heating rates with various benefits of the eco-friendly approach.[22]

Microwave-induced synthesis of aromatic polyamides by the phosphorylation reaction reported an optimization of the Yamazaki–Higashi method of

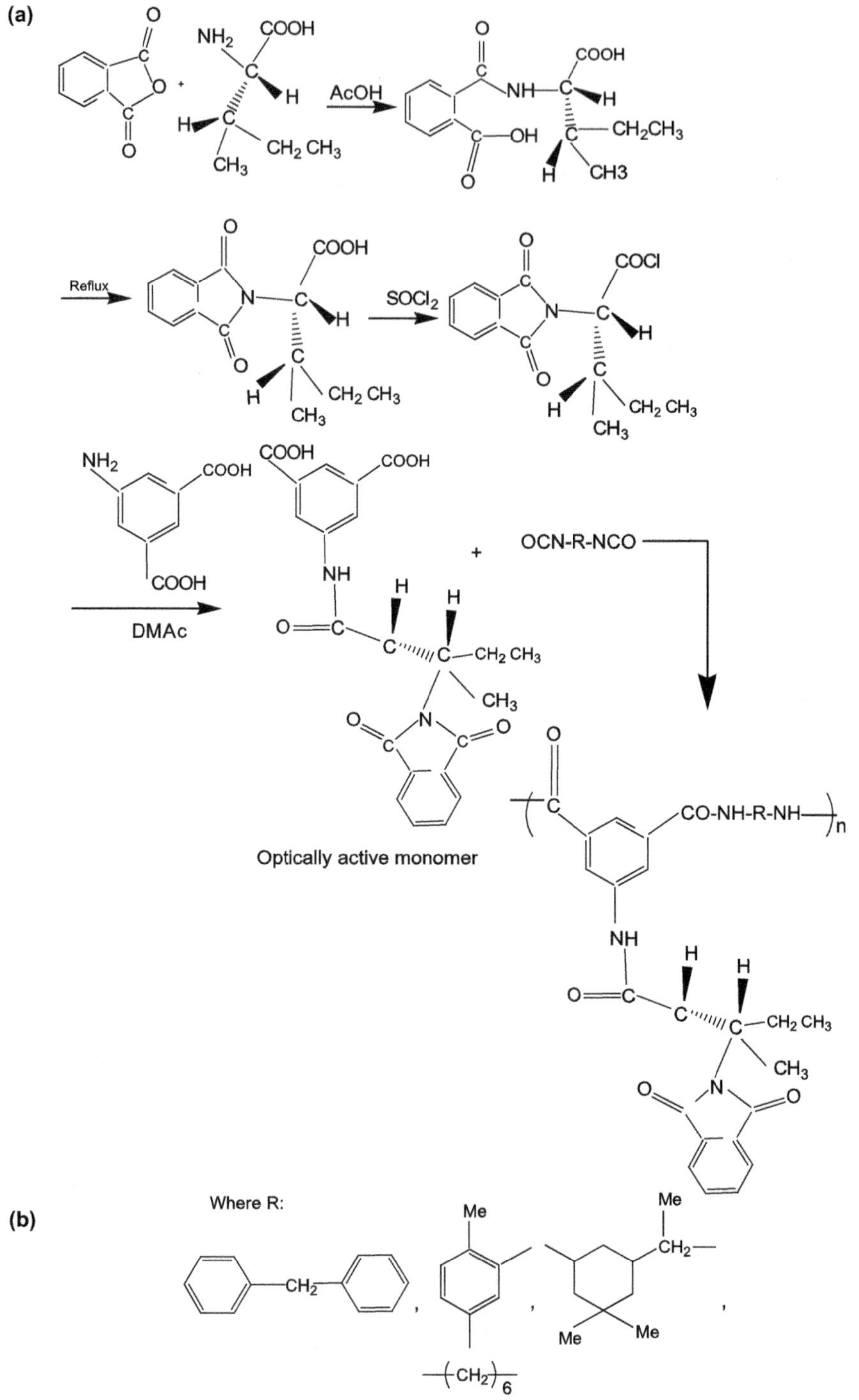

(a)

(b)

Scheme 3.4 Synthesis of optically active 5-(3-methyl2-phthalimidylpentanoylamino) isophthalic acid and its subsequent polymerization with different diisocynates.[22]

direct polyamidation from aromatic diacids and aromatic diamines. As a model reaction, the polycondensation between 4,4'-oxydianiline and isophthalic acid was studied in the presence of a phosphorylation agent and under microwave irradiation. This experimental study revealed that optimal results can be attained by using a great excess of pyridine as a catalyst at moderate radiation intensity (200 W), and up to 25% (w/v) monomer concentration. Although the polymer properties achieved by microwave and by conventional heating were comparable, an exceptional reaction rate under microwave irradiation yielded a polyamide inherent viscosity of around 1 dLg^{-1} in only a few minutes.[23]

3.3 Synthesis of Poly(imide)s

Due to their excellent thermal, mechanical and chemical stabilities, linear aromatic polyimides are often used in high-performance applications. Imai *et al.*[12] researched the fact that aliphatic polyimides can be successfully synthesized by the melt polycondensation of the nylon-salt-type monomers composed of aliphatic diamines and aromatic tetracarboxylic acids or their diester-diacids. They applied this finding to the polycondensation of the nylon-salt-type monomers under MW irradiation for the synthesis of polyimides (Scheme 3.5). Post this establishment, a large number of the polyimides containing the pyromellitoyl unit in their polymer chains were prepared by the step-growth polymerization of aliphatic diamines with both pyromellitic acid and its diethyl ester.[10]

The authors emphasized the necessity of polar organic medium for polymerization to proceed and used *N*-methylpyrrolidone, *N*-cyclohexyl-pyrrolidone, 1,3-dimethylirnidazolidon and tetramethylene sulfone as

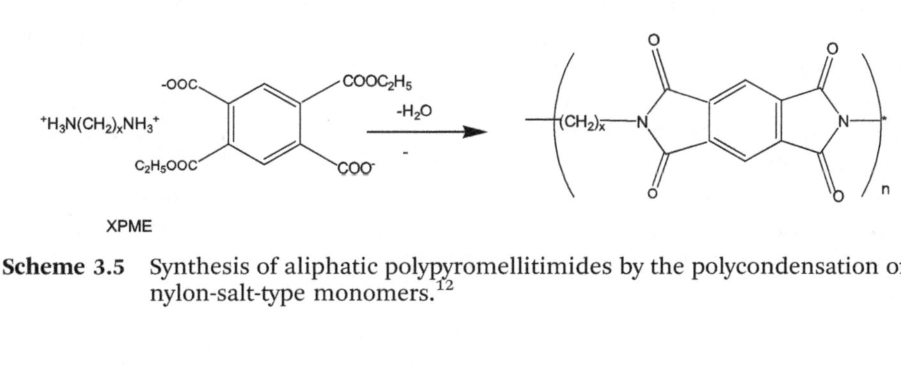

Scheme 3.5 Synthesis of aliphatic polypyromellitimides by the polycondensation of nylon-salt-type monomers.[12]

solvents. The comparative account showed that 1,3-dimethylirnidazolidon was best suited as a primary microwave absorber and solvent for the monomers and polymers. The authors obtained polyimides with inherent viscosities of 0.7 dL g^{-1} (for $x = 12$) within two minutes from pyromellitic acid, the starting material. Control experiments were set for comparison between MW irradiation and conventional heating for the reaction of dodecamethylene diamine with pyromellitic acid. The results showed that the solution polymerization under MW conditions proceeds much faster than the corresponding solid-state synthesis under conventional heating during the initial five minutes. However, it is interesting to note that the inherent viscosities of the polymers that were prepared in the MW oven only slightly increased with prolonged irradiation times, while those from the polymers that were synthesized with conventional heating still increased to values of up to 2 dL g^{-1}. No explanation was provided for these observations; instead, the authors emphasized the microwave's superiority to conventional heating during the first five minutes.

The synthesis of polyimides, having third order non-linear optical properties, from sodium tetrazodiphenyl naphthionate and pyromellitic dianhydride under microwave irradiation as well as oil-bath heating in N,N-dimethylformamide (DMF) as a solvent by a two-step method.[4] Using microwave heating, the imidization time was reduced from hours to minutes, and due to the fast heating rate, the imidization degree was noticeably increased. They also introduced polyimides[4] containing the pyromellitoyl unit *via* a two-step pathway. In the first step, from reaction of benzoguanamine and 2,4-tolylenediisocyanate generated polyurea, then polyimidation of the resulting polyurea and pyromellitic dianhydride in solid phase gave polyimide. They found that the degree of imidization under microwave irradiation reached a maximum value in 8 min, while by conventional heating it reached its highest value in 5 h. Li and co-workers reported the copolycondensation of aromatic dianhydrides and diamines by the direct polycondensation under microwave irradiation as well as conventional solution polycondensation *via* a two-step procedure.[24,25] A series of mono and di-imide compounds was synthesized by the reaction of common aromatic diamines with 4,5-dichlorophthalic acid in aqueous solution (at temperature between 160 °C and 200 °C) as a precursor to determining the chemical reactivity changes in these diamines during copolyimide synthesis under the same conditions. The reactivities of the second amino group were shown to reduce dramatically, in a number of examples, after substitution had occurred on the first amino group. The effects of these reactivity changes on polymer and copolymer properties were examined by the synthesis of a series of polymers containing two of the diamines with very different reactivity behaviors.[26] Microwave-assisted polycondensation of [4,4′-(hexafluoroisopropylidene) diphthalic anhydride, pyromellitic dianhydride] and [2,4,6-trimethyl-*m*-phenylenediamine] to produce polyimides was reported by Tellez *et al.*[27]

3.4 Poly(amide-imide)s

Due to their properties, like retention of good mechanical properties at high temperatures, easier processability on comparison with other aromatic thermostable polymers such as polyamides and polyimides, there is a growing interest in poly(amide-imide)s (PAIs). They are being applied in several areas like adhesives, electronic wire enamel, injection-molding, extrusion products and membranes.[28–30]

Faghihi *et al.* reported the synthesis and properties of new optically active PAIs from the polycondensation reactions of *N,N'*-(pyromellitoyl)-bis-l-phenylalanine diacid chloride (4) with six different derivatives of 5,5-disubstituted hydantoin compounds (a–f) as a heterocyles unit under MW irradiations.[31] They used a Samsung domestic microwave oven (2450 MHz, 900 W) for carrying out polycondensation reactions.

The mechanism (Scheme 3.6) involved the reaction of pyromellitic dianhydride (1,2,4,5-benzenetetracarboxylic acid 1,2,4,5-dianhydide) (1) reacted with l-phenylalanine (2) in a mixture of acetic acid and pyridine (3 : 2) at room temperature. *N,N'*-(pyromellitoyl)-bis-l-phenylalanine diacid (3) was obtained in quantitative yield on refluxing at 90–100 °C. The imide-acid (3) was converted to *N,N'*-(pyromellitoyl)-bis-l-phenylalanine diacid chloride (4) by reaction with thionyl chloride. Rapid and highly efficient synthesis of poly(amide-imide)s (6a–f) was achieved under MW irradiation. The reaction proceeded in the presence of a small amount of *o*-cresol, as a polar organic medium that acts as a primary microwave absorber. The polycondensation proceeded rapidly, compared with the conventional melt polycondensation and solution polycondensation, and was almost completed within 10 min, giving a series of poly(amide-imide)s with inherent viscosities about 0.28–0.44 dL/g. The resulting poly(amide-imide)s were obtained in high yield and were optically active and thermally stable. The authors studied thermal properties of the poly(amide-imide)s using thermal gravimetric analysis (TGA).[31] These optically active pyromellitoyl polymers were used as model substances for column materials in enantioselective chromatography.[31,32]

With a potential application as a packaging material in column chromatography, Mallakpour *et al.* prepared a library of poly(amide-imide)s from the reactions of optically active *N,N'*-(4,4'-carbonyldiphthaloyl)-bis-l-phenylalanine diacid chloride and the corresponding l-alanine and l-leucine compounds, with eight aromatic diamines, all of them with a rigid scaffold (Scheme 3.7).[33–35] The two monomers were dissolved in *o*-cresol and exposed to microwave irradiation for 7 to 10 min in a domestic microwave oven, to yield polymers with large inherent viscosities in the range of 0.22 to 0.85 dLg^{-1}.

The resulting polymers were characterized by FTIR, elemental analysis and specific rotation. The authors also reported a detailed thermal analysis of the polymers obtained. The authors presented a comparative account of microwave assisted polymerization in a Teflon vessel with that in a porcelain dish. They observed that, for similar reaction times, the polymers recovered

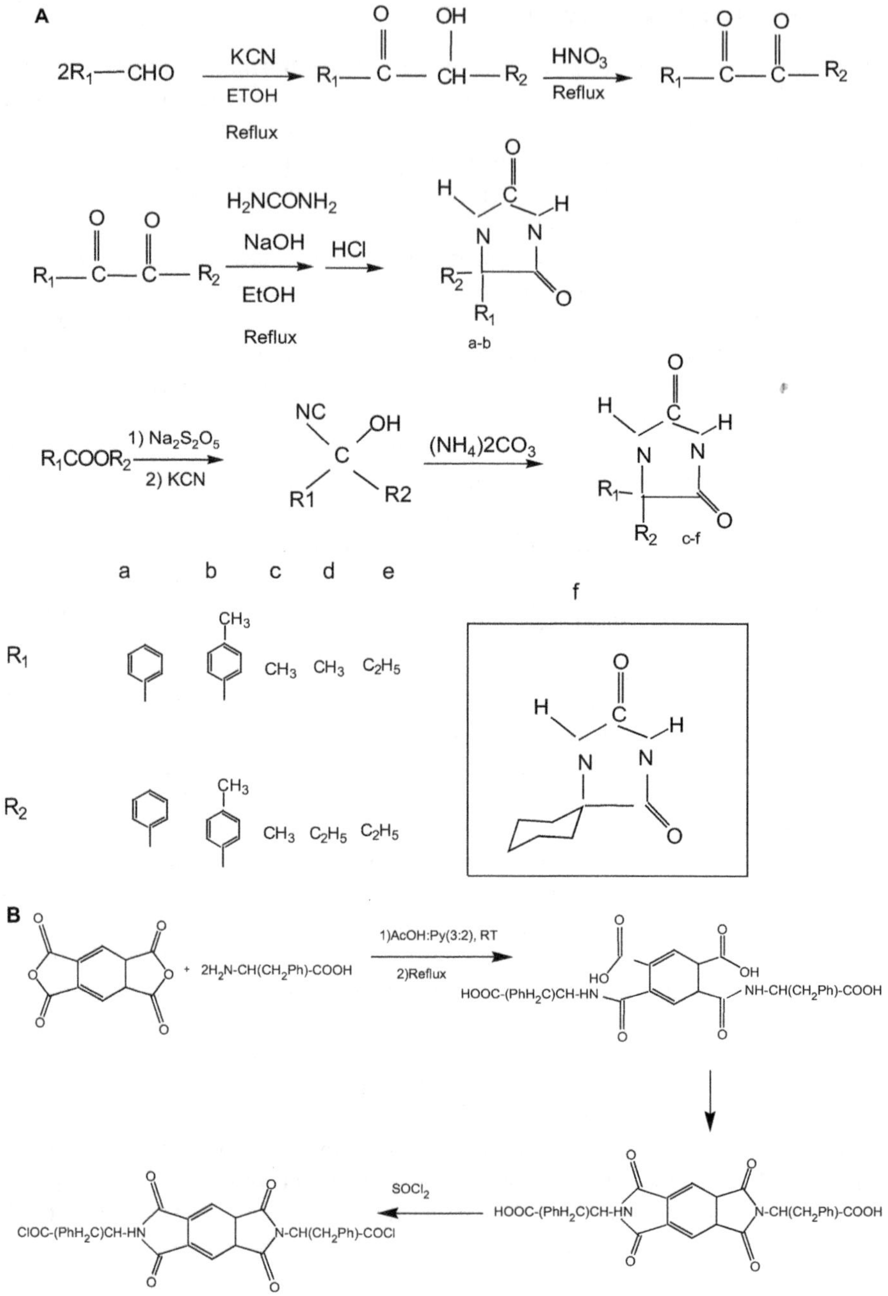

Scheme 3.6 Synthesis of Poly (amide imides) by microwave-assisted polycondensation of *N,N*-(Pyromellitoyl)-bis-l-phenylalanine diacid chloride with six different derivatives of 5,5-disubstituted hydantoin compounds (a–f). (A) Synthesis of 5,5-disubstiuted hydantoin derivative (a–f), (B) Synthesis of *N,N*-(Pyromellitoyl)-bis-l-phenylalanine diacid chloride, (C) Polymerization of the monomers.[31]

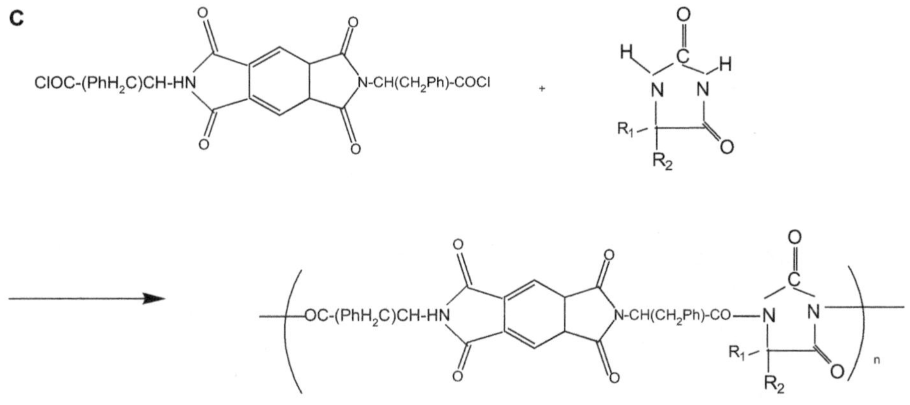

Scheme 3.6 (Continued)

from the porcelain dish had higher inherent viscosities.[33] Apart from this, for two series of compounds, the microwave-assisted polycondensation was compared with a solution polymerization supported by trimethylsilyl chloride (for the activation of thediamines).[33,34] The authors performed these reference experiments in nitrogen atmosphere, followed by a subsequent increase of the temperature to ambient conditions. From the properties of the polymers, it could be concluded that the microwave irradiation was superior to the catalyzed solution polymerization. In a more recent publication, these findings were repeatedly observed for the corresponding step-growth polymerizations of N,N'-(pyromellitoyl)-bis-L-isoleucine diacid chloride and six aromatic diamines.[36]

Applying the advantages of microwave irradiations to step-growth polymerization, polymerization of N,N'-(4,4′-carbonyldiphthaloyl)-bis-L-alanine diacid chloride with bulky derivatives of tetrahydropyrimidone and tetrahydro-2-thioxopyrimidine (six compounds) has been successfully carried out (Scheme 3.8).[37] Using o-cresol as a solvent and a MW absorber, the monomers were irradiated for 10 min in a domestic microwave oven to yield, (quasi) quantitatively, polymers with inherent viscosities of 0.25 to 0.45 dL g^{-1}. The polymers obtained (aromatic PAIs) were optically active and soluble in various organic solvents and possessed good thermal stability. These polymers have the potential to be used in a proper column chromatography technique for the separation of enantiomeric mixtures.

Faghihi *et al.* applied similar reaction conditions to obtain optically active poly(amide-imide)s from N,N'-(4,4′-carbonyldiphthaloyl)-bis-L-alanine diacid chloride and six representative hydantoin and thiohydantoin derivatives.[38] The polymerizations were carried out in solution in o-cresol in a domestic microwave oven to produce polymers within irradiation times of 10 min. Mallakpour *et al.* synthesized a similar set of compounds from N,N'-(4,4′-hexafluoroisopropylidene)-bis-phthaloyl-L-leucine diacid chloride and a group of ten aromatic diamines (Scheme 3.9); the polymers exhibited viscosities of 0.50 to 1.93 dL g^{-1}.[39]

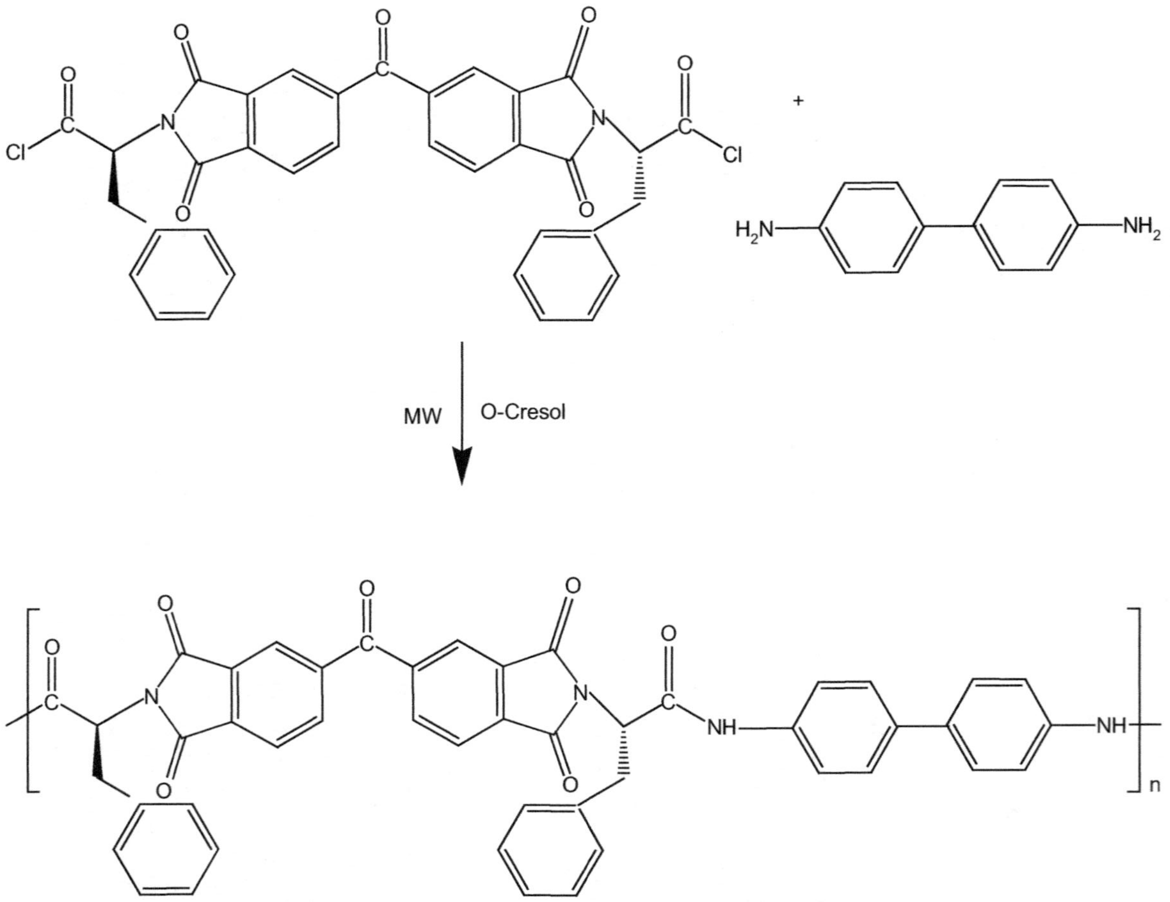

Scheme 3.7 Synthesis of poly(amide-imide)s from N,N'-(4,4′-carbonyldiphthaloyl)-bis-L-phenylalanine diacid chloride and L-alanine compounds.[33–35]

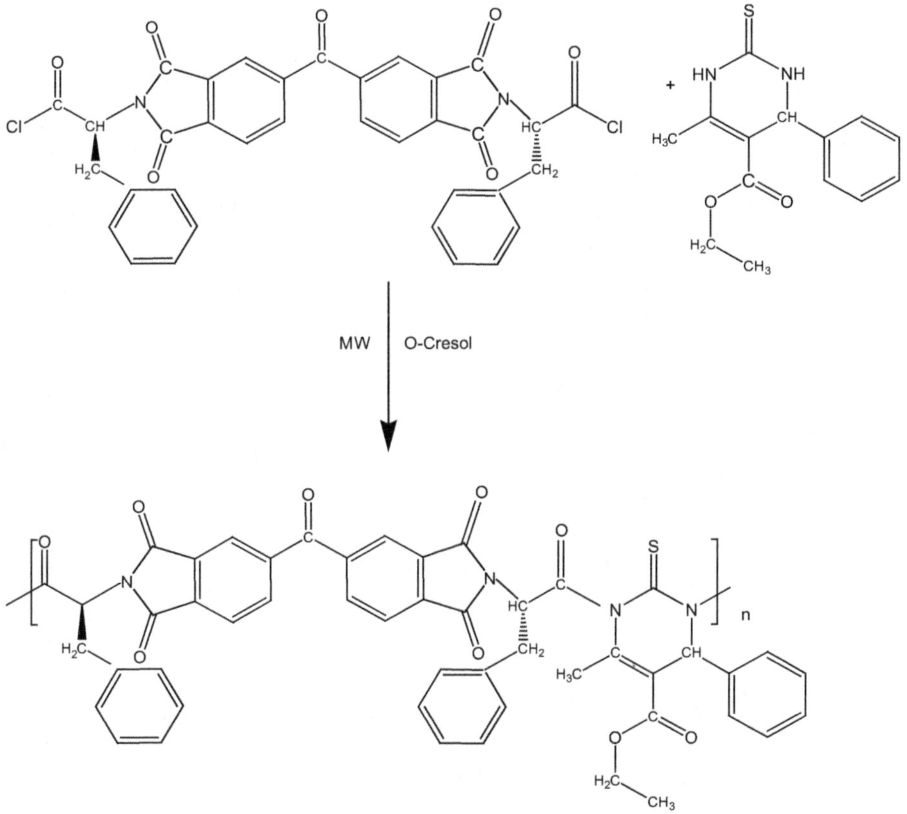

Scheme 3.8 Synthesis of poly(amide-imide)s from *N,N'*-(4,4'-carbonyldiphthaloyl)-bis-L-phenylalanine diacid chloride and the corresponding L-leucine compounds.[37]

Mallakpour *et al.* introduced the MW-assisted synthesis of optically active poly(amide-imide)s containing EPICLON [3a,4,5,7a-tetrahydro-7-methyl-5-(tetrahydro-2,5-dioxo-3-furanyl)-1,3-isobenzofurandione]or [5-(2,5-dioxotetrahydrofurfuryl)-3-methyl-3-cyclohexyl-1,2-dicarboxylic acid anhydride] as structural unit (Scheme 3.10). The process emphasized the universality of the concept for the microwave-assisted step-growth polymerization.[39,40] The authors carried out polymerizations themselves with seven different aromatic diamines using *N*-methylpyrrolidone as solvent in a domestic microwave oven. With exposure times of 5 min, poly(amide-imide)s with inherent viscosities of 0.12 to 0.22 dLg^{-1} were obtained. In order to probe and establish microwave's superiority over formerly exercised polymerization techniques, control experiments were set up: The low-temperature solution polycondensation (in *N*-methylpyrrolidone with trimethylsilyl chloride as promotor), as well as the high-temperature analogon, rendered polymers with comparable properties. These alternate reactions required

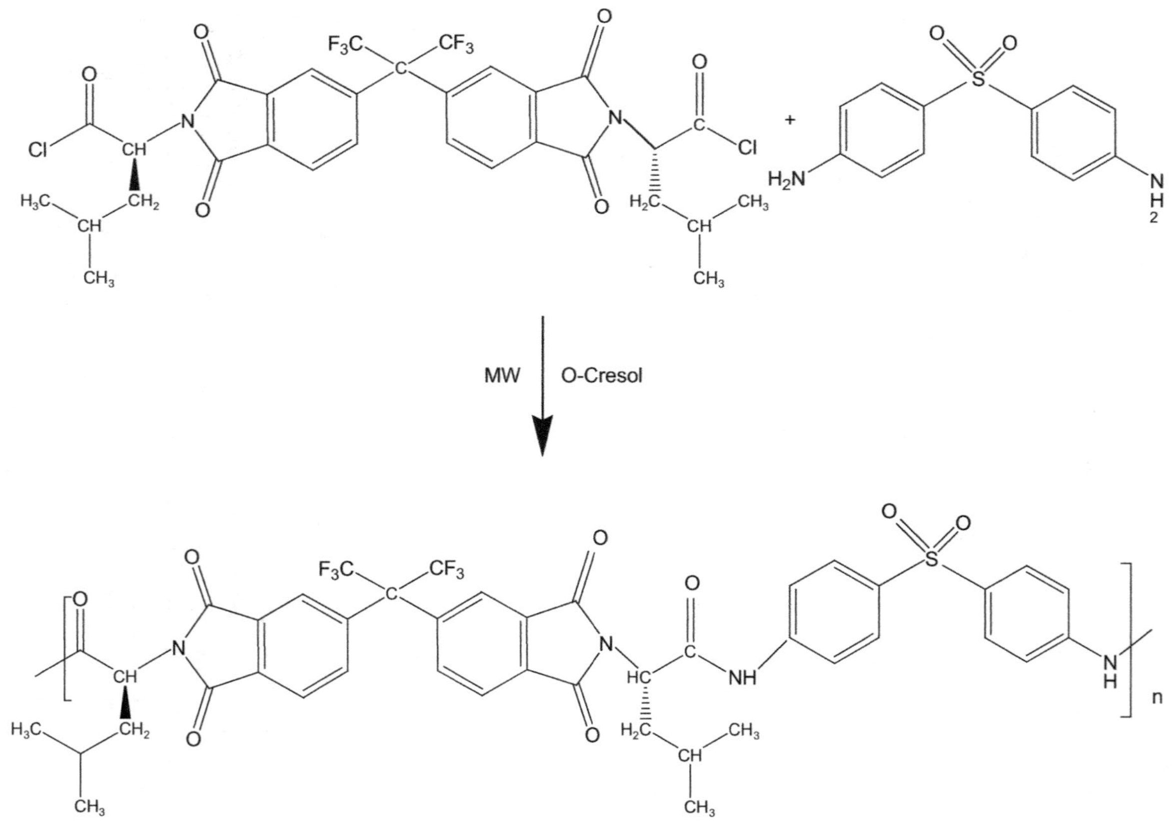

Scheme 3.9 Schematic representation of reaction N,N'-(4,4′-hexafluoroisopropylidene)-bis-phthaloyl-L-leucine diacid chloride and aromatic diamine.[39]

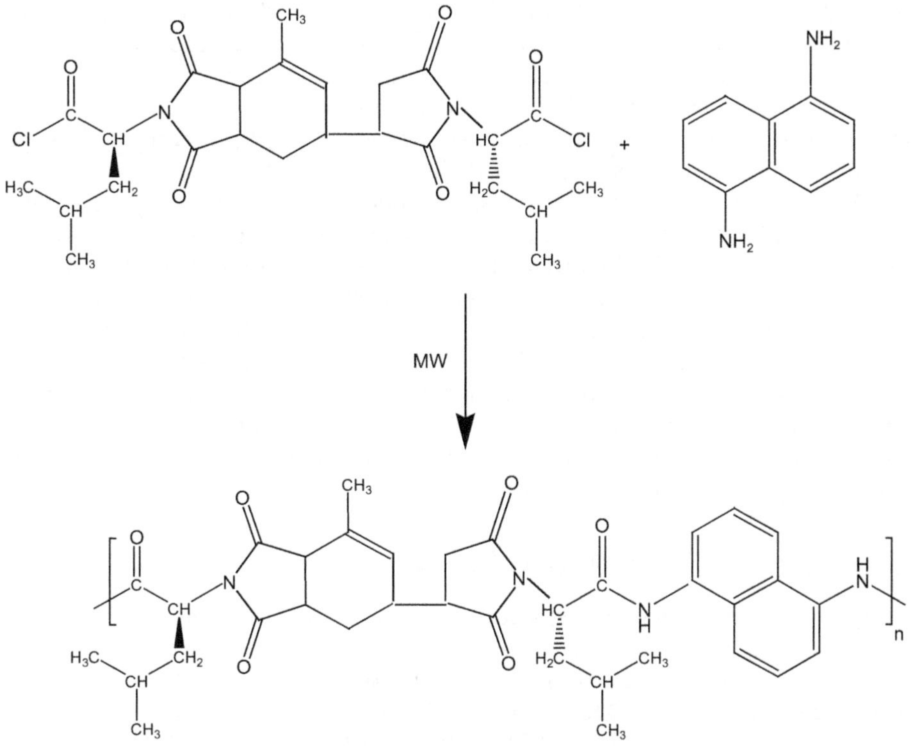

Scheme 3.10 here (chemical structures)

Scheme 3.10 Synthesis of Epiclon-containing optically active poly(amide-imide)s.[40]

significantly longer reaction times in the range of hours, in contrast to minutes for MW-assisted polymerizations.

In the synthesis of poly(amide-imide)s from N,N'-(4,4′-sulfonediphthaloyl)-bis-L-phenylalanine diacid chloride and its L-leucine congener with seven aromatic diamines, similar reaction trends were observed on changing the polymerization technique from conventional heating to MW-assisted in organic media.[41,42]

3.5 Synthesis of Poly(urea)s

Mallakpour *et al.* also introduced the step-growth polymerization concept to the synthesis of otherwise not-so-popular polyureas.[43–45] This non-popularity can be attributed to their high melting points and low solubility. However, due to the presence of –NH–CO–NH– functional group, they are capable of hydrogen bonding from different sides. Being polyamides of carbonic acid, they are tough, high melting and suitable for fiber applications.[46] The authors reacted 4-(4′-Aminophenyl)-1,2,4-triazolidine-3,5-dione 1 mol of acetyl chloride in dry N,N dimethylacetamide (DMAc) at 15 °C to obtain high yields of 4-(4′-acetamidophenyl)-1,2,4-triazolidine-3,5-dione [4-(4′-acet-anilido)-1,2,4-triazolidine-3,5-dione] (APTD). The reaction of the APTD monomer with excess *n*-isopropylisocyanate was performed at room temperature in DMAc solution. The group performed step-growth polymerization reactions of monomer APTD with hexamethylene diisocyanate, isophorone diisocyanate, and tolylene-2,4-diisocyanate under MW irradiation and solution polymerization in the presence of pyridine, triethylamine or dibutyltin dilaurate as a catalyst (Scheme 3.11).

MW-assisted polycondensation proceeded rapidly, as compared with conventional solution polycondensation. The resulting polyureas possessed an inherent viscosity in the range of 0.07–0.17 dL/g in dimethylformamide or sulfuric acid at 25 °C. These polyureas were characterized by IR, ^{1}H-NMR, elemental analysis and thermogravimetric analysis. The control set up comprised both step-like heating procedures as well as reflux conditions, both of them performed under conventional heating. Even with a broad variation of the catalysts utilized, it was found that microwave irradiation was superior to the other polymerization techniques, shortening reaction times from up to 24 h down to 8 min.

3.6   Synthesis of Poly(ether)s

Zsuga and co-workers reported a successful homopolymerization of D,L-lactic acid (2-hydroxypropanoic acid) in MW reactors.[47] Due to the high dipole moment of the monomer bulk polymerization with MW as heating source was highly successful. The authors obtained oligomers with molecular weights in the range of 600 to 1000 daltons within 20 min with MW heating (domestic microwave oven, 650 W) as compared to 24 h with conventional heating, exhibiting an acceleration factor of 70 for the transfer to the MW reactor. Using matrix-assisted laser-desorption ionization time-of-flight mass spectrometry (MALDI-TOF MS), the authors also reported that prolonged exposure times to MW irradiation induced an undesirable formation of cyclic oligomers.

Liu *et al.* investigated the polymerization of a derivative of lactic acid, namely L-2-hydroxy-3-phenylpropanoic acid (Scheme 3.12).[48] They carried out the polymerizations with variable time and power in a domestic MW oven. With maximum power (510 W), polymers with number-average molecular weights of 1 800/3 900/5 400 and PDI values of 1.8/1.0/1.4 were obtained within 0.5/1.5/2.5 h, respectively.

Polyanhydrides being biodegradable polymers represent potential materials for drug delivery applications. Polyanhydrides are usually prepared by a two-step process under conventional heating. The steps comprise (a) the synthesis of a polyanhydride prepolymer, its isolation, purification and (b) subsequent polymerization. Mallapragada and co-workers showed that the polymerization can be carried out in a single step, admittedly with intermediate removal of unreacted acetic anhydride (Scheme 3.13).[49]

The reaction proceeded by placing dicarboxylic acid (like sebacic acid) and excessive acetic anhydride in a high-pressure capped vial and irradiated for 2 min in a microwave oven (General Electrics, 1100 W). Decapping the vial

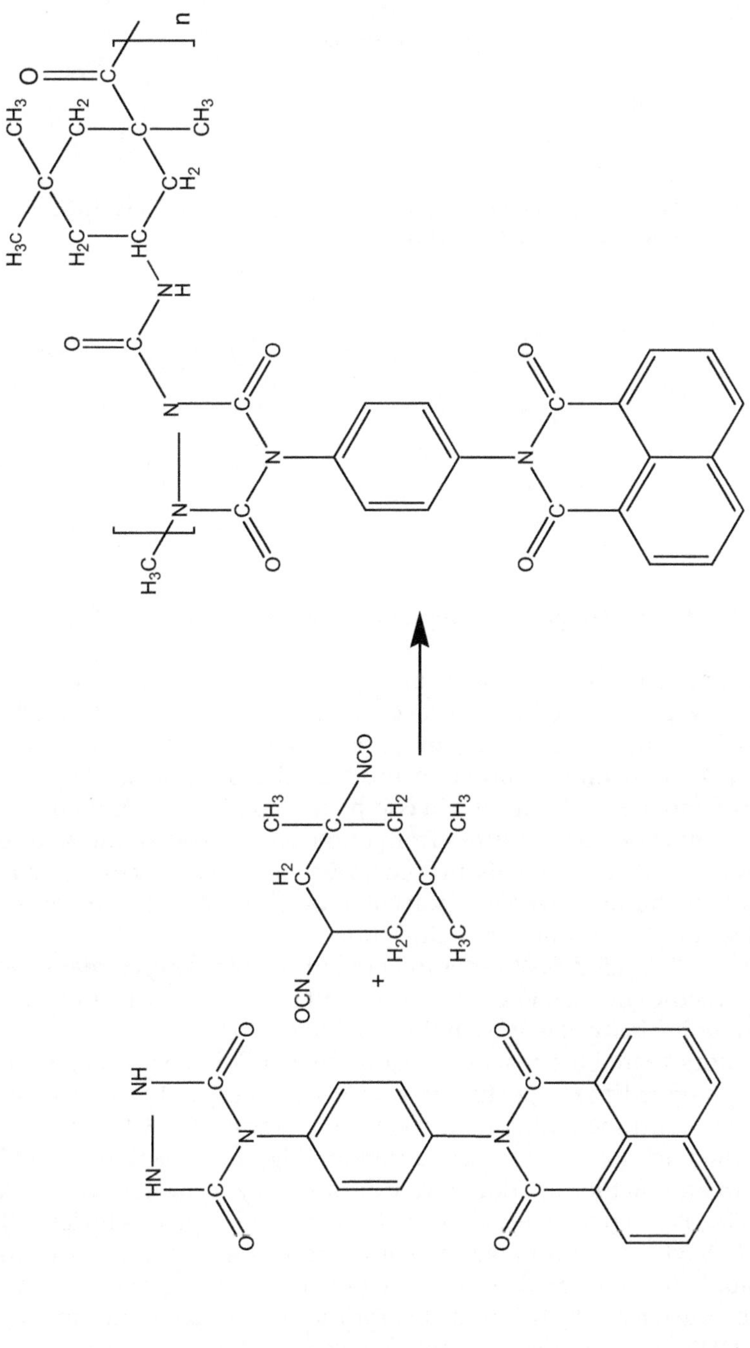

Scheme 3.11 Application of step-growth polymerization to the synthesis of polyureas.[43]

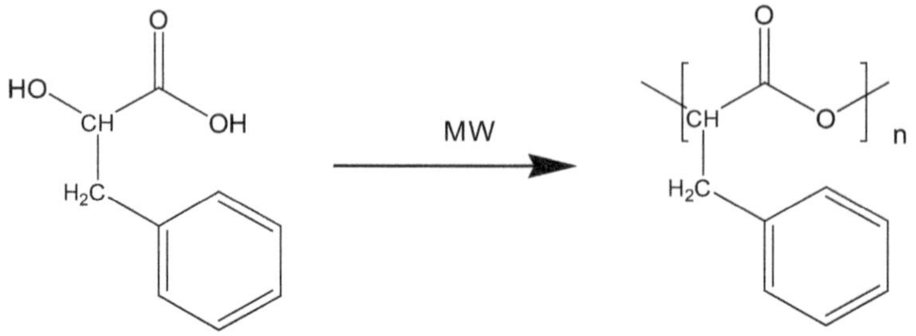

Scheme 3.12 MW-assisted polymerization of L-2-hydroxy-3-phenylpropanoic acid (a derivative of lactic acid).[48]

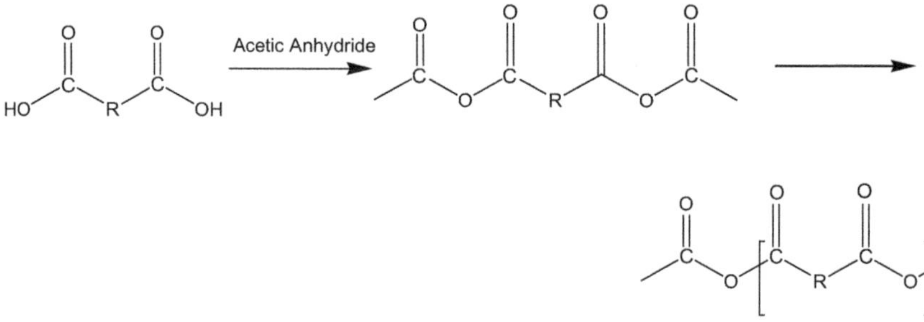

Scheme 3.13 Single step polymerization of polyanhydride polymer.[49]

and purging the hot solution with inert gas immediately after the reaction led to the evaporation of the unreacted acetic anhydride. After eventual addition of a catalyst (SiO_2, Al_2O_3, or glass beads), the vial was recapped and heated for 4 to 25 min in order to produce the polyanhydride. The step economy due to the elimination of a complete step of the reaction procedure led to shortening of reaction times from four days to a few minutes. Amongst the catalysts used, glass beads proved to be the most successful support. Polymers with number-average molecular weights of 11 400 daltons were obtained within 25 min of irradiation time.

Alimi *et al.*[50] applied MW radiations to study homo-polycondensation of 1-chloro-4-methoxylbenzene in solution in alkaline dimethyl sulphoxide for producing poly(phenylene vinylene)-ether (PPVether).

The authors heated the reaction mixture to 200 °C in a modified domestic MW oven, stopped the exposure as soon as the targeted temperature was reached and continued stirring at room temperature for 6 h. The polymer was obtained in 43% yield; two different oligomer fractions soluble in chloroform and dichloromethane were additionally recovered with yields of 52 and 5%. This copolymer, derived from poly-phenylene-vinylene (PPV), denoted PPV–ether has tunable optical properties, is amorphous in nature and insoluble in common solvents. The polymer found application in LEDs and was characterized by infrared absorption (IR), Raman scattering, optical density (OD), gravimetric thermal analysis (GTA), differential thermal analysis (DTA), X-ray photoelectron spectroscopy (XPS) and photo-luminescence. MW-assisted polycondensation of isosorbide or isoiodide with 1, 8-dibromo- or 1,8-dimesyl-octane in the presence of a small amount of toluene under phase-transfer catalytic (PTC) conditions within 30 min has also been reported.[17] In the case of isosorbide, microwave-assisted polymerization showed increased total yields compared with conventional heating.

Gerardin *et al.* utilized MW irradiation to synthesize polyglycerols from glycerol carbonate.[51] The authors concluded that under MW conditions (Scheme 3.14), the polymerization could be successfully carried out at a faster rate and without any solvent.

On comparison with classic alkaline polymerization directly from glycerol, the authors mentioned that though it is a more efficient procedure with high atom economy, the MW-assisted polymerization of glycerol carbonate has advantages like the use of less-hazardous conditions, better energy efficiency, and safer solvent. Also, it does not require any solvent reaction or purification procedure.

Compared to classical thermal heating, MW irradiation is characterized by a fast and homogeneous heating pattern, which allows high-temperature chemistry. The authors also emphasized the nonthermal microwave effects due to specific heating of polar intermediates produced during the reaction, leading to modified selectivity enabling polymerization that cannot be performed with thermal heating.[52]

3.7 Poly(ether imide)s and Poly(ester imide)s

A significant increase in reaction rates for the imidization reaction between poly(tetramethylene oxide)glycol di-*p*-aminobenzoate and benzene-tetra-carboxylic acid dianhydride was observed by Yu and co-workers.[53] On changing the heating mode from a conventional oil bath to a MW reactor the reaction times were reduced from 12 to 3 h and the reaction temperatures were decreased from 115 to 60 °C. The water molecules released during the imidization reaction acted as good absorbers of the MW irradiation and gave an additional edge to MW reactions.

Mallakpour *et al.* in an analogy to the synthesis of optically active poly-(amide imide)s, prepared poly(ester imide)s with chiral carbon atoms in the polymer chain containing pyromellitoyl units.[54,55] The authors synthesized a basic library of 14 optically active poly(ester imides). The compounds *N,N'*-(pyromellitoyl)-bis-L-phenylalanine diacid chloride and *N,N'*-(pyromellitoyl)-bis-L-leucine diacid chloride were synthesized with the help of conventional heating. Post synthesis of the diacid chlorides, the step-growth poly-merizations involving these and a series of aromatic diols with rigid scaf-folds (like phenolphthalein, bisphenol A, 4,4-hydroquinone, and others) was carried out in a domestic MW oven. Taking into consideration the low dipole moments and the corresponding low absorbance of the educts, these poly-merization reactions were carried out in *o*-cresol, where *o*-cresol doubles up

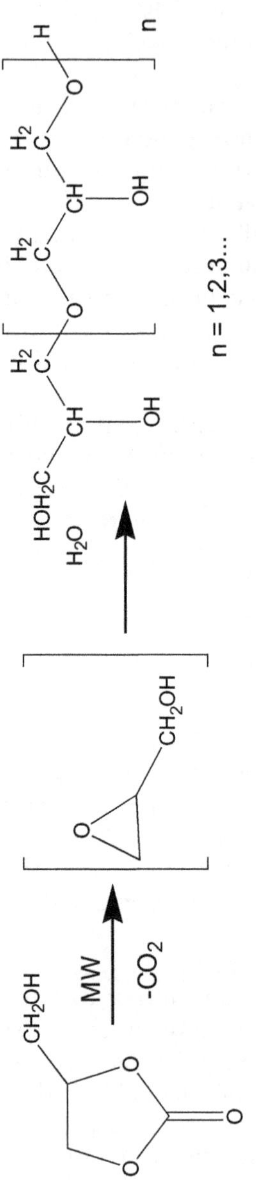

Scheme 3.14 Microwave-assisted polymerization of glycerol carbonate.[51]

as solvent and as primary MW absorber at the same time; 1,4-diazabicyclo[2.2.2]octane was used as catalyst. Within reaction times of 10 min, polymers with inherent viscosities in the range of 0.10 to 0.27 dL/g were readily obtained. Comparison experiments under solution conditions using triethylammonium chloride as a phase-transfer catalyst only produced oligomers, as was concluded from the improved solubility in methanol.

Another study comprised the synthesis of a 14-membered library with optically active carbonyldiphthaloyl anhydride derivatives, namely *N,N'*-(4,4'-carbonyldiphthaloyl)-bis-L-phenylalanine diacid chloride as well as the corresponding L-alanine compound, and a set of aromatic diols with rigid scaffolds.[56,57] The authors pointed out an interesting contrast in that the polymerizations involving the L-phenylalanine derivative[56] could not be performed under MW irradiation, but only at low temperatures under phase-transfer catalysis. With the L-alanine compound,[58] on the other hand, a microwave-assisted synthesis was successfully carried out in *o*-cresol without the addition of catalysts. Polymers with inherent viscosities of 0.35 to 0.58 dL/g were obtained within 12 min in a domestic microwave oven.

A similar study focused on the reaction of 4,4'-(hexafluoroisopropylidene)-*N,N'*-bis-(phthaloyl-L-leucine) diacid chloride with aromatic diols, all of them with rigid fixation of the hydroxy function.[59] The authors performed a systematic investigation using three different solvents *viz. o*-cresol, *m*-cresol and 1,2-dichlorobenzene, with a bifunctional role as solvent and primary microwave absorber. *O*-cresol proved to be best suited for the above-stated purposes. Reactions in a domestic MW oven were completed within five minutes and yielded polymers with inherent viscosities of 0.50 to 1.12 dL/g. The conventional heating was carried out in 1,2-dichlorobenzene, only yielded polymers with inherent viscosities of 0.09 to 0.18 dL/g, even after 24 h reaction time. The corresponding polymers derived from *N,N'*-[4,4'-carbonyl-bis-(phthaloylimido)]-bis-L-leucine diacid chloride, however, were found not to be formed under MW irradiation, but from low-temperature polycondensation in chloroform with triethylamine as catalyst. This establishes the efficacy of MW as a superior heating source.[60]

3.8 Polymerizations Involving C–C Coupling Reactions

MW-assisted polymerizations by C–C coupling reactions are not very popular reactions and have gained attention only very recently contrary to other very well-established organic synthesis reactions.[61–67] Apart from the C–C coupling reactions described hereafter, so far only one metathesis polymerization of phenylacetylene under MW irradiation has been reported. The peculiarity of the reaction was *in situ* generation of catalyst ([(arene)M(CO)$_3$] complexes (M = Cr, Mo, W)). These catalysts reduced the reaction time from 24 h (under reflux conditions of a solution in 1,2-dichloroethane) to 1 h (in a domestic MW oven).[68]

Coupling polymerization of 2,7-dibromo-9,9-dihexylfluorene using nickel(0) as catalyst and toluene as solvent was investigated by Carter.[69] The reaction under conventional heating required an activation step for the preparation of the catalyst. The polymerization reaction was a slow, tedious process that lasted up to 24 h. The polymerization showed a low reproducibility and a hindered access to high molecular-weight polymers which was attributed to the decreasing solubility of the polymers with an increasing degree of polymerization. The reaction under MW irradiation was carried out in capped vials in a customized MW reactor specially designed for chemical synthesis (SmithCreator, PersonalChemistry), taking place in high-temperature and high-pressure synthesis. The polymers were obtained in almost quantitative yield after 10 min at 250 °C. The MW irradiations improvised the reaction in multiple ways, which included:

✓ Tremendous decrease in reaction times.
✓ Minimum side reactions.
✓ Single step polymerization with no prior activation of the catalyst in a separate procedure.
✓ High molecular weight polymers (100 000 daltons or even higher).

The molecular weights could be controlled by the addition of a monofunctional end-capping unit (4-bromobiphenyl) to yield polymers with molecular weights from 5000 to 40 000 daltons and PDI values from 1.65 to 2.22, respectively. The reaction is depicted in the following (Scheme 3.15).

Yamamoto and co-workers described the preparation of poly(pyrazine-2,5-diyl) from 2,5-dibromopyrazine in a CEM Discover apparatus.[70] They used the same nickel(0)-based catalyst system, and observed a similar shortening of reaction times down to 10 min.

Barbarella and co-workers prepared thiophene oligomers using the Suzuki coupling reaction under MW conditions.[71] They used Synthewave 402 apparatus(Prolabo) for the synthesis of quinquethiophenes, from bulk conditions of 2-thiophene boronic acid and dibromo precursors with three thiophene units, catalyzed and promoted by [PdCl$_2$(dppf)], KF, and KOH. (Scheme 3.16). The reaction was completed in 10 min and attained a maximum temperature of 70 °C with the yield of 74%. The synthesis of polymers instead of oligomers, however, was not carried out. The reaction is depicted in the following Scheme 3.16.

Leadbeater *et al.* prepared biaryl compounds by Suzuki coupling reactions in a microwave reactor.[72] Scherf and co-workers were first to apply Suzuki and Stille cross-coupling reactions for the preparation of semiconducting polymers in a monomodal MW reactor (CEM Discovery).[73] They prepared five representative polymers with number-average molecular weights in the range of 11 300 to 15 400 daltons in a few minutes under MW irradiation, which required three days of conventional heating. The authors varied the applied power to optimize reaction temperatures for each specific reaction.

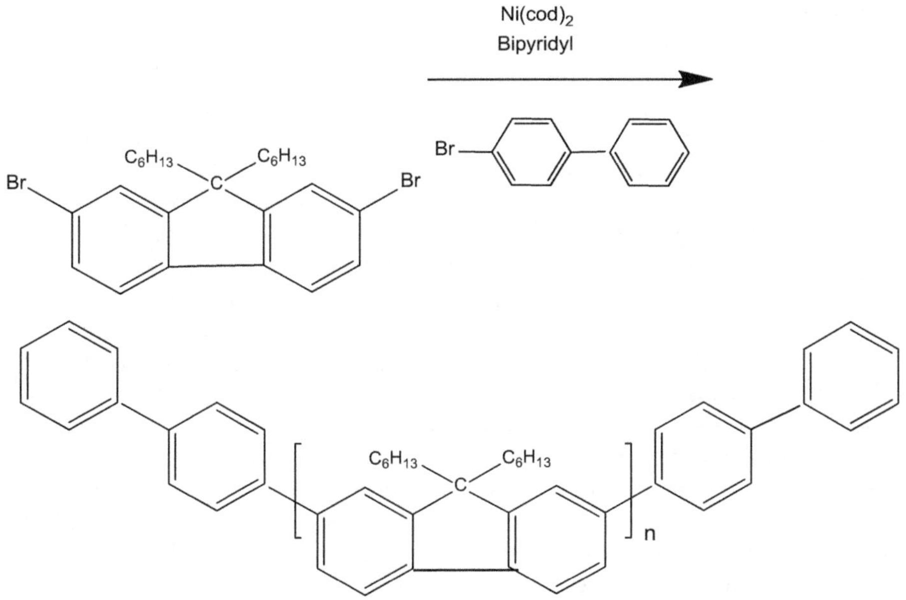

Scheme 3.15 Coupling polymerization of 2,7-dibromo-9,9-dihexylfluorene using nickel(0) as catalyst.[69]

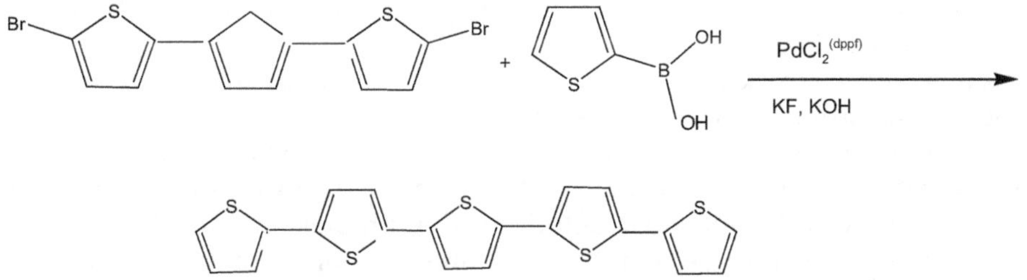

Scheme 3.16 Preparation of thiophene oligomers using the Suzuki coupling reaction under MW conditions.[71]

The MW irradiations were particularly effective in the case of the Stille cross-coupling polymerization between the electron rich (and consequently less active) 1,5-dioctyloxy-2,6-dibromonaphthalene and 5,5′-bis(trimethyl-stannyl)-2,2′-bithiophene which yielded polymers with higher molecular weights (13 700 compared with 5100 daltons, the latter under conventional heating).

3.9 Phase-transfer Catalysis

MW-assisted phase transfer catalysis has been applied mainly to the synthesis of polyethers. Hurduc *et al.* were amongst the first to study step-growth polymerization of 3,3-bis(chloromethyl)oxetane and several bisphenols using phase transfer catalysis.[74] The authors prepared the targeted compounds from a water/nitrobenzene two-phase system with tetrabutylammonium bromide as phase-transfer catalyst. For comparative analysis they carried out the reaction both with conventional heating and MW assistance in a domestic MW oven.

It was observed that on being compared with conventional heating for 5 h, the same polymer yields could be obtained under MW irradiation in the temperature range of 95 to 100 °C (depending on the bisphenol's chemical structure) within 90 min. The authors attributed the accelerated reaction rates to the enhanced reaction temperatures. The thermal analysis of the polymer revealed higher glass temperatures and melting points for the polymers synthesized in the microwave oven, which are indicative of higher molecular weights. Baudel *et al.* published a similar study comprising MW-assisted polymerizations of 3,3-bis(chloromethyl) oxetane and several bisphenols, performed in a monomodal microwave system (Synthewave 402, Prolabo) and compared with results from conventional heating. The difference in reaction times, 4 to 6 h in the case of conventional heating as opposed to 20 min under MW irradiation (for the preparation of oligomers with oligomers' degrees of 6 to 8 and comparable yields in the range from 64 to 97%), was found to be very pronounced.[75] Zhang and co-workers related an increase in reaction speed of the phase-transfer catalyzed step-growth polymerization of bisphenol A and bis(chlorophthalimide) in a domestic microwave oven to the selective excitation of the phenol anion which in turn led to enhanced reactivity.[76]

Loupy and co-workers studied shifts in selectivity for the polyetherification of isosorbide (1,4:3,6-dianhydro-D-sorbitol) or isoidide (1,4:3,6-dianhydro-L-iditol) with 1,8-dibromo and 1,8-dimesyloctane on carrying out under MW irradiations (Scheme 3.17).[77,78] They performed a phase-transfer catalyzed polymerization on a solution of equimolar amounts of the corresponding educts in toluene in the presence of tetrabutylammonium bromide and aqueous potassium hydroxide Synthewave 402, Prolabo as a MW reactor. For the isosorbide-containing polymers, the authors observed that the reaction rates accelerated under MW irradiation, providing polymers in yields of around 70% within 30 min. The same reaction required 24 h for similar completion in conventional heating. From gel permeation chromatography (GPC) and MALDITOF MS analysis it was concluded that the polymers synthesized in the microwave reactor showed higher molecular-weight averages. Isoidide derivative, interestingly, showed comparable yields under both MW irradiation and conventional heating, but a strong tendency to higher polymerization degrees under microwave assistance was observed, as a heavy weight fraction (insoluble in methanol) was formed in relatively higher yield (39–67% compared with 5–12%). Besides, the two different modes of activation showed specific mechanisms of chain termination: the polyethers prepared under conventional heating had hydroxylated ends, and those synthesized in the microwave reactor exhibited

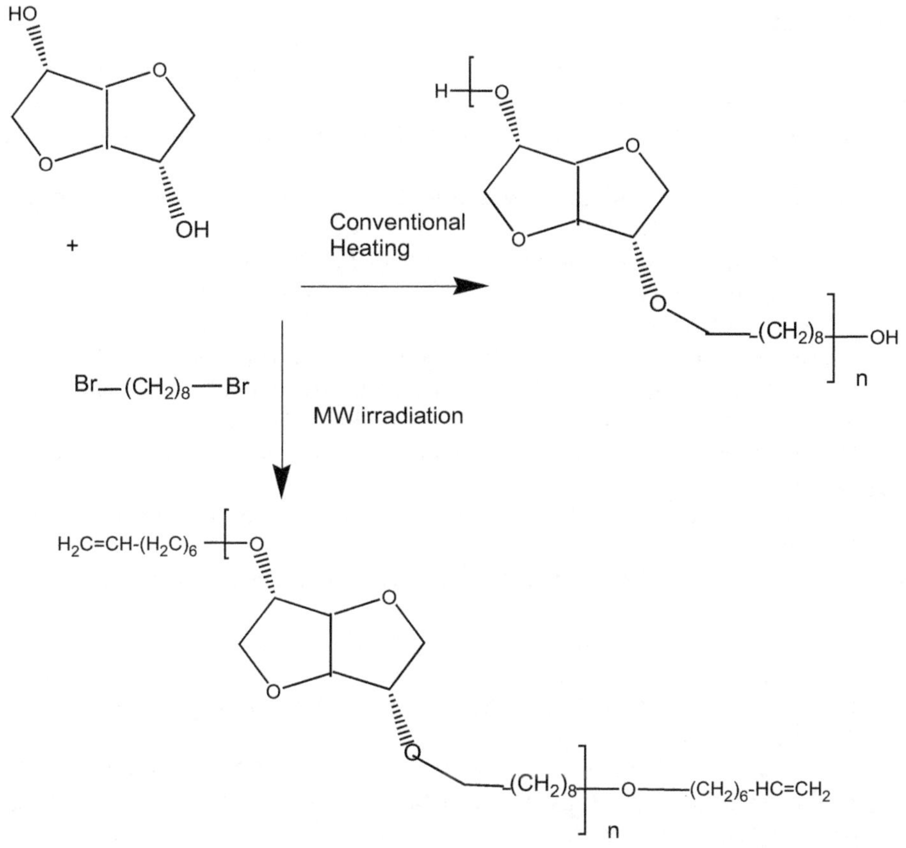

Scheme 3.17 Step-growth polymerization of isoidide with 1,8-dibromooctane and the different mechanisms of chain termination, depending on the activation source.[79]

ethylenic group ends (Scheme 3.17). These characteristics were also observed in the phase-transfer catalyzed step-growth polymerizations of 1,8-dibromo- and 1,8-dimesyloctane with isosorbide derivatives (two isosorbides linked by alkyl or glycol chains, possessing two equivalent alcohol functions in exo positions).[79]

A recent review paper on MW-assisted step-growth polymerization by Durka *et al.* has tried to shed light on the role of important process parameters, such as the presence and type of solvent, the dielectric properties of the mixture and the individual phases, the use of heterogeneous catalysts, pressure, stirring, reflux conditions, temperature measurement method and microwave absorbing fillers.[80]

Application of microwave heating to step-growth reactions has witnessed a spectacular growth during the last 15 years. However, no protocol has been developed for such reactions yet. Experiments have been carried out at laboratory scale using widely different experimental procedures.

References

1. Paul J. Flory, *Principles of Polymer Chemistry*, Cornell University Press, Ithaca, USA, 1953, p. 39.
2. R. Hoogenboom and U. S. Schubert, *Macromol. Rapid Commun.*, 2007, **28**, 368.
3. F. Wiesbrock, R. Hoogenboom and U. S. Schubert, *Macromol. Rapid Commun.*, 2004, **25**, 1739.
4. S. Mallakpour and Z. Rafiee, *Iran. Polym. J.*, 2008, **17**, 907.
5. R. Hoogenboom, T. F. A. Wilms, T. Erdmenger and U. S. Schubert, *Aust. J. Chem.*, 2009, **62**, 236.
6. S. Mallakpour and Z. Rafiee, *Prog. Polym. Sci.*, 2011, **36**(12), 1754.
7. J. Cheng, J. Zhou, Y. Li, J. Liu and K. Cen, *Energy Fuels*, 2008, **22**, 2422.
8. S. Sinnwell and H. Ritter, *Aust. J. Chem.*, 2007, **60**, 729.
9. H. Yanagawa, K. Kojima, M. Ito and N. Handa, *J. Mol. Evol.*, 1990, **31**, 180.
10. Y. Imai, H. Nemoto, S. Watanabe and M. A. Kakimoto, *Polym. J.*, 1996, **28**, 256.
11. Y. Imai, *Polym. Prepr. (Am. Chem. Soc. Div. Polym. Chem.)*, 1995, **36**, 711.
12. S. Watanabe, K. Hayama, K. H. Park, M. Kakimoto and Y. Imai, *Macromol. Chem. Rapid Commun.*, 1993, **14**, 481.
13. K. H. Park, S. Watanabe, M. Kakimoto and Y. Imai, *Polym. J.*, 1993, **25**, 209.
14. F. K. Mallon and W. H. Ray, *J. Appl. Polym. Sci.*, 1998, **69**, 1203.
15. A. Pourjavadi, M. R. Zamanlu and M. J. Zohuriaan-Mehr, *Angew. Makromol. Chem.*, 1999, **269**, 54.
16. K. Faghihi and M. Hagibeygi, *Eur. Polym. J.*, 2003, **39**, 2307.
17. *Microwave in Organic Synthesis*, ed. A. Loupy, Wiley-VCH Verlag GmbH, Weinheim, 2002.
18. A. Caouthar, A. Loupy, M. Bortolussi, J. C. Blais, L. Dubreucq and A. Meddour, *J. Polym. Sci., A Polym. Chem.*, 2005, **43**, 6480.
19. S. Mallakpour and Z. Rafiee, *Polym. Adv. Technol.*, 2008, **19**, 1474.
20. S. Mallakpour and M. Taghavi, *J. Appl. Polym. Sci.*, 2008, **109**, 3603.
21. S. Mallakpour and A. Zadenhnzari, *Amino Acids*, 2009, **38**, 1369.
22. S. Mallakpour and M. Dinari, *J. Appl. Polym. Sci.*, 2009, **112**, 244.
23. A. C. Paula, A. S. Ricardo, B. R. Mercier, E. L. A. Angel, G. C. A. Jose and A. J. Abajo, *Aust. J. Chem.*, 2009, **62**, 250.
24. Q. Li, Z. Xu and C. Yi, *J. Appl. Polym. Sci.*, 2008, **107**, 797.
25. Q. Li, X. Yang, W. Chen, C. Yi and Z. Xu, *Macromol. Symp.*, 2008, **261**, 148.
26. B. Dao, A. M. Groth and J. Hodgkin, *Eur. Polym. J.*, 2009, **45**, 1607.
27. H. M. Tellez, J. Alquisira, J. G. Lopez-Cortes and C. Alvarez-Toledano, *J. Appl.Polym. Sci.*, 2010, **116**, 2816.
28. P. E. Cassidy, *Thermally Stable Polymers, Synthesis and Properties*, Dekker, New York, 1980.

29. A. V. R. Reddy, P. S. Reddy and S. Anand, *Eur. Polym. J.*, 1998, **34**, 1441.
30. D. Fritsch and K. V. Peinemann, *J. Membr. Sci.*, 1995, **99**, 29.
31. K. Faghihi, K. Zamani, A. Mirsamie and S. Mallakpour, *J. Appl. Polym. Sci.*, 2004, **91**, 516.
32. S. E. Mallakpour, A. Hajipour and S. Habibi, *Eur. Polym. J.*, 2001, **37**, 2435.
33. S. E. Mallakpour, A. Hajipour and M. R. Zamanlou, *J. Polym. Sci., Part A: Polym. Chem.*, 2001, **39**, 177.
34. S. E. Mallakpour, A. Hajipour and M. R. Zamanlou, *Eur. Polym. J.*, 2002, **38**, 475.
35. S. E. Mallakpour, A. Hajipour and K. Faghihi, *Eur. Polym. J.*, 2001, **37**, 119.
36. S. E. Mallakpour and M. H. Shahmohammadi, *J. Appl. Polym. Sci.*, 2004, **92**, 951.
37. S. E. Mallakpour, A. Hajipour, K. Faghihi, N. Foroughifar and J. Bagheri, *J. Appl. Polym. Sci.*, 2001, **80**, 2416.
38. K. Faghihi, K. Zamani, A. Mirsamie and M. R. Sangi, *Eur. Polym. J.*, 2003, **39**, 247.
39. S. E. Mallakpour, A. Hajipour and S. Khoee, *J. Polym. Sci., Part A: Polym. Chem.*, 2000, **38**, 1154.
40. S. E. Mallakpour, A. Hajipour and M. R. Zamanlou, *J. Polym. Sci., Part A: Polym. Chem.*, 2003, **41**, 1077.
41. S. E. Mallakpour and M. R. Zamanlou, *J. Appl. Polym. Sci.*, 2004, **91**, 3281.
42. S. E. Mallakpour and E. Kowsari, *J. Polym. Sci., Part A: Polym. Chem.*, 2003, **41**, 3974.
43. S. E. Mallakpour and E. Kowsari, *J. Appl. Polym. Sci.*, 2004, **91**, 2992.
44. S. E. Mallakpour and Z. Rafiee, *J. Appl. Polym. Sci.*, 2004, **91**, 2103.
45. S. E. Mallakpour and Z. Rafiee, *J. Appl. Polym. Sci.*, 2003, **90**, 2861.
46. P. W. Morgan, *Condensation Polymers: By Interfacial and Solution Methods*, Wiley-Interscience, New York, 1965, ch. 5.
47. S. Keki, I. Bodnar, J. Borda, G. Deak and M. Zsuga, *Macromol. Rapid Commun.*, 2001, **22**, 1063.
48. L. Liu, L. Q. Liao, C. Zhang, F. He and R. X. Zhuo, *Chin. Chem. Lett.*, 2001, **12**, 761.
49. B. M. Vogel, S. K. Mallapragada and B. Narasimhan, *Macromol. Rapid Commun.*, 2004, **25**, 330.
50. K. Alimi, M. Molinie, J. C. Majdoub, J. L. Bernede, H. Fave, Bouchriha and M. Ghedira, *Eur. Polym. J.*, 2001, **37**, 781.
51. K. Iaych, S. Dumarcay1, E. Fredon, C. Gerardin, A. Lemor and Gerardin, *J. Appl. Polym. Sci.*, 2011, **120**, 2354.
52. S. Martinez-Gallegos, M. Herrero and V. Rives, *J. Appl. Polym. Sci.*, 2008, **109**, 1388.
53. J. Chen, Q. Chen and X. Yu, *J. Appl. Polym. Sci.*, 1996, **62**, 2135.
54. S. E. Mallakpour and S. Habibi, *Eur. Polym. J.*, 2003, **39**, 1823.

55. S. E. Mallakpour, A. Hajipour and S. Habibi, *J. Appl. Polym. Sci.*, 2002, **86**, 2211.
56. S. E. Mallakpour, A. Hajipour and M. Zamanlou, *J. Appl. Polym. Sci.*, 2002, **85**, 315.
57. S. E. Mallakpour, A. Hajipour and K. Faghihi, *Polym. Int.*, 2000, **49**, 1383.
58. S. E. Mallakpour, A. Hajipour and S. Khoee, *J. Appl. Polym. Sci.*, 2000, **77**, 3003.
59. S. E. Mallakpour, A. Hajipour and M. Zamanlou, *Polym. Sci., Ser. A*, 2002, **44**, 243.
60. A. Stadler and C. O. Kappe, *J. Comb. Chem.*, 2001, **3**, 624.
61. H.-K. Lee and T. M. Rana, *J. Comb. Chem.*, 2004, **6**, 504.
62. M. D. Evans, J. Ring, A. Schoen, A. Bell, Edwards, D. Berthelot, R. Nicewonger and C. M. Baldino, *Tetrahedron Lett.*, 2003, **44**, 9337.
63. N. S. Wilson, C. R. Sarko and G. Roth, *Tetrahedron Lett.*, 2001, **42**, 8939.
64. P. Lidström, J. Westman and A. Lewis, *Comb. Chem. High Throughput Screening*, 2002, **5**, 441.
65. W.-M. Dai, D.-S. Guo, L.- Sun and X.-H. Huang, *Org. Lett.*, 2003, **5**, 2919.
66. A. Finaru, A. Berthault, T. Besson, G. Guillaumet and S. Berteina-Raboin, *Org. Lett.*, 2002, **4**, 2613.
67. A. M. L. Hoel and J. Nielsen, *Tetrahedron Lett.*, 1999, **40**, 3941.
68. K. Dhanalakshmi and G. Sundararajan, *Polym. Bull.*, 1997, **39**, 333.
69. K. R. Carter, *Macromolecules*, 2002, **35**, 6757.
70. T. Yamamoto, Y. Fujiwara, H. Fukumoto, Y. Nakamura, S. Koshihara and T. Ishikawa, *Polymer*, 2003, **44**, 4487.
71. M. Melucci, G. Barbarella and G. Sotgiu, *J. Org. Chem.*, 2002, **67**, 8877.
72. N. E. Leadbeater and M. Marco, *Angew. Chem., Int. Ed.*, 2003, **42**, 1407.
73. B. S. Nehls, U. Asawapirom, S. Füldner, E. Preis, T. Farrell and U. Scherf, *Adv. Funct. Mater.*, 2004, **14**, 352.
74. N. Hurduc, D. Abdelylah, J.-M. Buisine, Decock and G. Surpateanu, *Eur. Polym. J.*, 1997, **33**, 187.
75. V. Baudel, F. Cazier, Woisel and G. Surpateanu, *Eur. Polym. J.*, 2002, **38**, 615.
76. C. Gao, S. Zhang, L. Gao and M. Ding, *J. Appl. Polym. Sci.*, 2004, **92**, 2415.
77. S. Chatti, M. Bortolussi, A. Loupy, J. C. Blais, D. Bogdal and M. Majdoub, *Eur. Polym. J.*, 2002, **38**, 1851.
78. S. Chatti, M. Bortolussi, A. Loupy, J. C. Blais, D. Bogdal and Roger, *J. Appl. Polym. Sci.*, 2003, **90**, 1255.
79. S. Chatti, M. Bortolussi, D. Bogdal, J. C. Blais and A. Loupy, *Eur. Polym. J.*, 2004, **40**, 561.
80. M. Komorowska-Durka, G. Dimitrakis, D. Bogdał, A. I. Stankiewicz and G. D. Stefanidis, *Chem. Eng. J*, 2015, **264**, 633.

CHAPTER 4

Ring-opening Polymerization

4.1 Introduction

Ring-opening polymerization (ROP) reactions are a type of chain-growth polymerization, in which the terminal end of a polymer acts as a reactive center, where further cyclic monomers join to form a larger polymer chain through ionic propagation. The treatment of some cyclic compounds with catalysts brings about cleavage of the ring followed by polymerization to yield high-molecular-weight polymers. ROP is a type of polymerization in which a cyclic monomer yields a monomeric unit that is either acyclic or contains fewer cycles than the monomer. As in chain polymerization, ROP also consists of a sequence of initiation, propagation and termination. When the reactive center of the propagating chain is a carbocation, the polymerization is called cationic ring-opening polymerization. When the active center is a carbanion, the reaction is an anionic ring-opening polymerization, *e.g.* polymerization of polypropylene oxide.

A different type of ring-opening polymerization, based on olefin metathesis, uses catalysts rather than cationic or anionic propagation.[1] Figure 4.1 gives the schematic representation of the ROP.

The ROP technique is used for polymerization of a variety of cyclic monomers, such as cyclic ethers, acetals, amides and esters siloxanes. These polymers represent an area of great interest for academic, industrial and biomedical concerns.[2]

The ROP technique is a relatively slow method when compared to polymerizations of monomers containing a carbon–carbon double bond (such as vinyl), the chain propagation rate constants for ROP are several orders of magnitude lower.[3] The ROP was found to benefit significantly from microwave-assisted heating, allowing a good control over the polymerizations combined with a tremendous acceleration of polymerization rates. However, the potential of microwave heating for the ROP of caprolactone was

RSC Green Chemistry No. 35
Microwave-Assisted Polymerization
By Anuradha Mishra, Tanvi Vats and James H. Clark
© Anuradha Mishra, Tanvi Vats and James H. Clark 2016
Published by the Royal Society of Chemistry, www.rsc.org

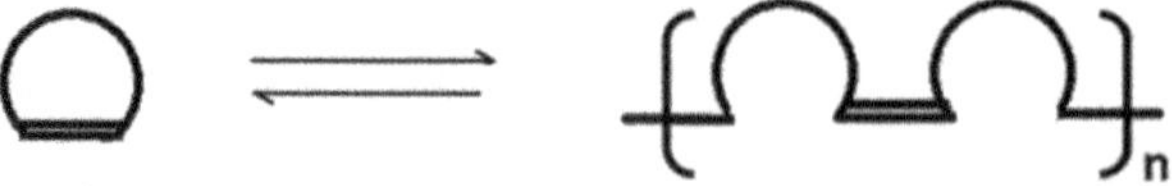

Figure 4.1 Schematic representation of ring-opening polymerization.

controversial initially as indicated in reports on thermal as well non thermal effects.[4] The controversy has motivated researchers to deepen their work in the area of microwave-assisted ROP in recent years.

4.2 Ring-opening Polymerization of Cyclic Esters

Cyclic esters such as lactones and lactides result in the formation of biodegradable aliphatic polyesters. These materials are promising candidates for bio-related applications such as drug-delivery or as scaffolds in tissue engineering. The majority of investigations on microwave-assisted ring-opening polymerizations of cyclic esters were performed using ε-caprolactone or lactides. Because of their biocompatibility, thermal properties and potential chemical functionalization, these polymers have become a center of research interest. In the past decade, several publications on kinetic studies of the metal-catalyzed ROP of ε-caprolactone under MW conditions have appeared[6] in scientific literature and most of them describe a significant reduction in reaction time for polymerizations performed under MW conditions.[5]

4.2.1 Ring-opening Polymerization of ε-caprolactone

Poly(ε-caprolactone) (PCL) has many applications in the biomedical field. It has been approved by the Food and Drug Administration (FDA) for use in the human body as a drug delivery device, suture (sold under the brand name Monocryl) and as scaffold for tissue engineering.[7] It also serves as an alternative for packing materials by virtue of having interesting properties such as miscibility with other polymers, besides being nontoxic, biodegradable and biocompatible.[8]

PCL can be synthesized by various methods. A few commonly used methods are:[9–13]

- Enzymatic synthesis.
- Step condensation polymerization.
- Monomer insertion mechanisms.
- Activated monomer mechanism.
- Use of alkoxides.

For biomedical applications, ultra-pure PCL is required. ROP of ε-caprolactone (CL) in supercritical CO_2 is reported to be a pollution-free mode of PCL preparation.[14] The microwave-assisted ring-opening polymerization of ε-caprolactone, (Scheme 4.1) exploiting a home-built monomodal microwave reactor, was first reported by Albert *et al.* The assembly used by them is schematically presented in Figure 4.2. They used pulsed microwave irradiations as a heat source and successfully polymerized CL with titanium tetrabutylate as a catalyst. A comparison of conversion, molecular weight and reaction kinetics of the microwave-assisted polymerization with thermal polymerization showed that the increased rate of reaction did not amount to a microwave effect as the changes were minor and within the experimental error.[15]

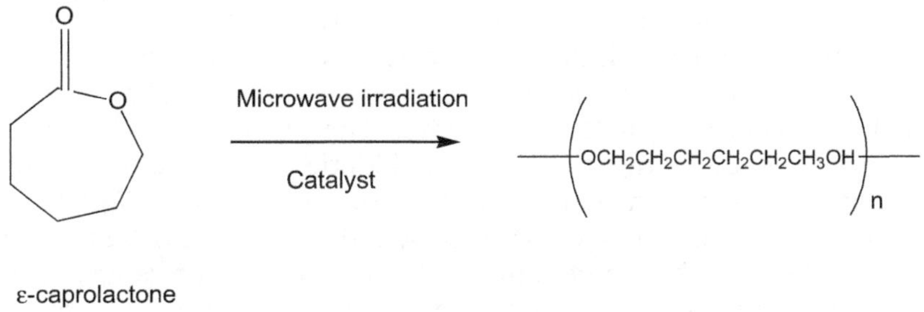

Scheme 4.1 Microwave-assisted ring-opening polymerization of ε-caprolactone.[15]

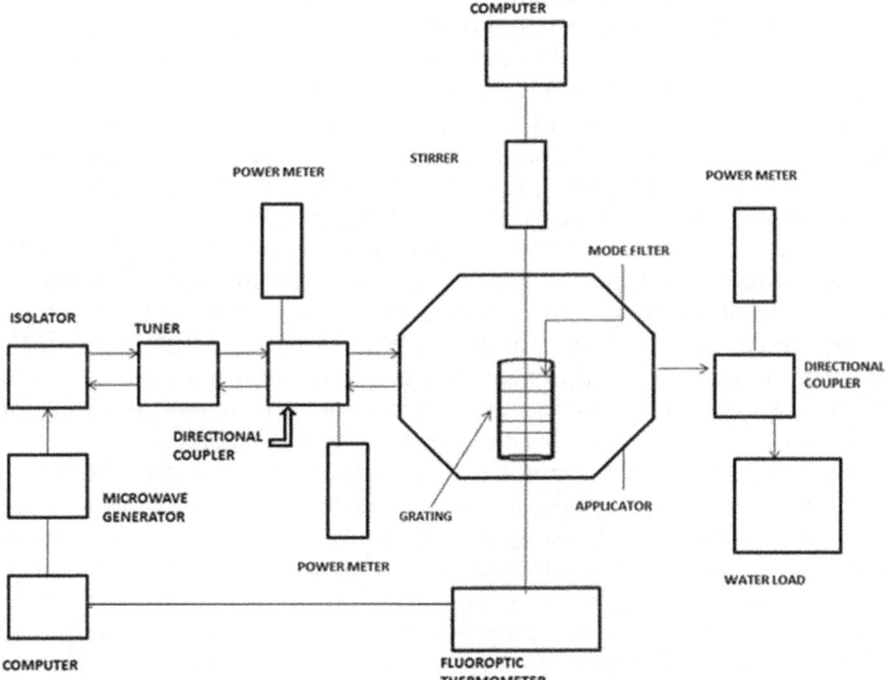

Figure 4.2 Schematic diagram of microwave reactor system used by Albert *et al.*[15]

It was not until 1998 that Fang and Scola[16] established the supremacy of microwave irradiation over conventional heating in cyclic ROP. They applied variable frequency (2.4–7.0 GHz) microwave irradiation to gain optimum processing control. They were able to attain temperatures between 150–200 °C under 70–100 W of microwave irradiation, to obtain PCL with a M_w of 9900–86 000 g mol^{-1} after 2 h of irradiation. The resulting products were characterized and it was confirmed that the quality of the resulting PCL (in terms of equivalent glass transition temperature (T_g), melting temperature (T_m) and thermal stability) produced by 2 h of microwave irradiation was akin to that produced by more than 12 h of the commercial thermal process.

Over time the standardization of microwave irradiated ROP started taking shape. Both academic and industrial researchers[17] tried to tame the process in terms of power and frequency of microwave operations, temperatures attained and control of molecular weights. The process was characterized by several tools including thermal and kinetic studies. Liu and Liao[18] conducted ROP of CL in a domestic microwave oven at a frequency of 2.45 GHz, under constant microwave powers of 170, 340, 510, and 680 W. They concluded that the temperatures of the polymerization ranged from 80 to 210 °C. PCL with a M_w of 124 000 g mol^{-1} and a yield of 90% was obtained at 680 W for 30 min using 0.1% (mol/mol) stannous octanoate (Sn(Oct)$_2$) as a catalyst. On irradiating the reaction at the same power for 270 mins using 1% (wt/wt) zinc powder as catalyst, the M_w of PCL was 92 300 gmol^{-1}.

While studying the thermal properties of microwave irradiated ring-opening of CL, Liu and Liao[18] found that the reaction temperature of CL was self-regulated to an equilibrium temperature. At 680 W of microwave irradiation, the temperature of CL (10 g) increased rapidly from 20 to 355 °C during the first 10 min and then fluctuated slightly around 360 °C during the next 20 min. An exothermic peak was observed from the thermogram of metal catalyzed ROP (MCROP) of CL with Sn(Oct)$_2$ as a catalyst. When a mixture of 10 g of CL with 0.1% (mol/mol) Sn(Oct)$_2$ was irradiated at 680 W, an exothermic peak appeared between 2 min and 5 min with a maximum temperature of 343 °C at 3 min. During this period of time (2–5 min), the MCROP of CL was very fast, resulting in PCL with a M_w of 123 000 g mol^{-1} and yield of 95%. The authors concluded that the thermal effect of microwave energy on the CL monomer and the reaction mixture was dependent on the power levels of microwaves and the mass scale of materials.

Sivalingam and Madras[6] studied the kinetics of MCROP of CL in bulk at 350 W with different cycle-heating periods (30–50 s). They set up two models for both thermal and microwave heating processes. The molecular weight distributions measured by GPC were determined as a function of reaction time. Because the temperature of the system continuously varied with the reaction time, a model was proposed based on continuous distribution kinetics with time/temperature-dependent rate coefficients. Experiments were conducted under thermal heating to quantify the effect of the microwave on polymerization. The polymerization also was investigated with thermal and microwave heating in the presence of a zinc catalyst. The activation energies determined from temperature-dependent rate coefficients for pure thermal heating, thermally aided catalytic polymerization, and microwave-aided catalytic polymerization were 24.3, 13.4, and 5.7 kcal/mol, respectively. This indicates that microwaves increase the polymerization rate by lowering the activation energy.

Sinnwell and co-workers[19] studied the direct synthesis of methacrylate PCL macromonomers. The MCROP of CL in the presence of methacrylic acid or acrylic acid yielded radical polymerizable polyester macromonomers. The main advantage of the process is the single step reaction. The thermal properties show that the melting point of the macromonomers was adjustable between 46 and 51 °C.

MW-assisted ROP can be benefited a lot by developing an appropriate catalyst. Owing to its application in the biomedical arena it is important to generate PCL by methods with the least residual reagents or biologically safe residues. The various catalysts used for catalytic polymerization of CL are zinc,[20] tin,[21] aluminum,[22] alkali metals,[23] early transition metals and rare earth halides,[22,24] and organo-lanthanide series.[25] Sivalingam *et al.* presented an elaborate study on the usage of zinc as a ROP catalyst. One of the major advantages of using zinc is its permissibility in the human body. Apart from this, the process is heterogeneous and separation is relatively easy.[6] However, high molecular weights are obtained by using coordinate catalysts.[22,24] Kowalski *et al.*[26] investigated the kinetics of the polymerization of CL initiated by stannous octanoate [$Sn(Oct)_2$] in tetrahydrofuran at 80 °C; the results strongly suggest that the tin alcoholate active centers were formed from the starting carboxylate, and the polymerization proceeded on the tin–alkoxide bond. Storey and Taylor[27,28] investigated ethylene glycol-initiated vacuum polymerization of CL, and the effect of $Sn(Oct)_2$ concentration on composition, molar mass and molar mass distribution was comprehensively examined. Barakat *et al.*[20] revealed that zinc alkoxides were effective initiators for living ROP of CL under very mild conditions in toluene solution, and PCL with a narrow molar mass distribution was obtained. Zinc halides were successfully used as initiators, and the M_w of PCL could be controlled by adjusting the monomer/initiator ratio.[29–31] The rare earth coordinate catalysts and one-component rare-earth catalysts also exhibited high activities for the polymerization of CL. Shen *et al.*[24] indicated that the catalyst activities of rare earth halides were greatly enhanced by epoxide.

Liu and co-workers[32] reported carboxylic acid catalyzed MCROP of CL. Benzoic acid and chlorinated acetic acids were used in the metal-free reaction. They took a mixture of CL and benzoic acid with a molar ratio of 25/1, the M_w and PDI of the PCL synthesized by microwave irradiation at 680W for 4 h were 44 800 g mol^{-1} and 1.6, respectively. Starting from the same mixtures, the M_w and PDI of the PCL synthesized by conventional heating at 210 °C for 4 h were 12 100 g mol^{-1} and 4.2, respectively. This clearly implied that the polymerization improved significantly under microwave irradiation both in terms of M_w and PDI. On initiating the reaction in the presence of chlorinated acetic acids, degradation of the resultant PCL was observed, which caused a reduction in M_W of PCL.

The effect of microwave energy on the chain propagation of benzoic acid-initiated MCROP of CL was also investigated.[33] The molar ratios of CL to benzoic acid used were 5, 15 and 25. On irradiating the mixtures under microwave, at powers ranging from 340 to 680 W, it was found that temperature was self-regulated to an equilibrated temperature between 204 and 240 °C. The chain propagated rapidly between 160 and 230 °C, however, degradation of the resultant PCL occurred when the temperature was above 230 °C. The results indicated that under microwave irradiation, there was significant enhancement of the propagation of PCL chains but at the same time the formation of growing centers at the beginning stage of the polymerization was prevented. PCL with a M_W over 4×10^4 g mol^{-1} was prepared with this metal-free approach.

Liu *et al.* studied acid-initiated MCROP of CL in context of its application in the preparation of a drug controlled release system.[34–36] The effects of microwave power, exposure time, monomer/catalyst molar ratio, and acidity of acid on the polymerization were the key parameters to be investigated. Both the polymerization rate and the M_W of polymers obtained were enhanced in comparison with the conventional thermal method. PCL with a M_W higher than 12 000 g mol^{-1} and a PDI below 1.6 was synthesized on using maleic acid, succinic acid and adipic acid. They also carried out the polymerization after mixing CL with a certain amount of ibuprofen (IBU), from which the IBU-PCL controlled-release system was prepared directly. The system was modeled on the basis of surface studies by scanning electron microscopy. The sustained release of IBU from the system from 12 h to 9 days with an increasing IBU loading level from 5 to 20 wt% presented a promising method to prepare drug controlled-release systems.

Lanthanide compounds are efficient, non-toxic catalysts for ROP of cyclic esters. MCROP of CL using early lanthanide halides as catalysts was investigated, using a monomodal reactor (Synthewave S402, Prolabo) and a multimode microwave oven.[37,38] Hydrated lanthanide halide catalyzed MCROP of CL is a very easy and efficient method to obtain macro-monomers with controlled M_W in the range of 3000–5000 g mol^{-1}, which can be used as starting materials for functionalization and copolymerization reactions (Scheme 4.2).

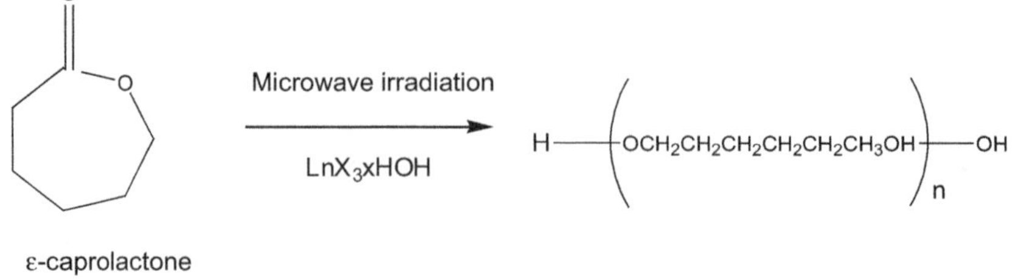

Scheme 4.2 Lanthanide halide catalyzed microwave-assisted ring-opening polymerization of ε-caprolactone.[38]

Nguyen *et al.*[39] have critically analyzed the acceleration in the ring-opening of lactones delivered by microwave heating. They reported that selective heating of the organometallic species in the ROP of CL contributes directly to differences in the reaction conditions and which need to be taken into account in comparisons between conventionally done and MW-assisted reactions.

Gong *et al.* reported 65% yield within 30 min in a MW-assisted ZnO-catalyzed ROP of ε-caprolactone in the presence of ionic liquid. MW-assisted ROP of ε-caprolactone in the presence of hydrogen phosphonate was reported to produce poly(ε-caprolactone) with molecular weight of 8100 g mol^{-1} in only 100 min.[40] Because of the strong absorption of the ionic liquid, high heating rates and equilibrium temperatures were obtained, which reduced the reaction time drastically. Besides the aforementioned advantages, based on the solubility difference between polymers and ionic liquids, polymers can be isolated by simple water extraction, which is considered a convenient and "environmentally friendly" process, while the ionic liquid can be efficiently recovered from water, thereby completely avoiding the use of volatile organic compounds during the purification step. These properties make the microwave-assisted ionic liquid initiated reactions a win–win combination in green chemistry.

4.2.2 Ring-opening Polymerization of Lactides

Poly(lactic acid) (PLA) is a biodegradable, biocompatible plastic and has potential application in the fields of biomedicines, agriculture and the packaging industry. Not only do properties such as excellent drug permeability and good mechanical strength make it suitable for the above-mentioned fields, its easy availability from inexpensive raw materials such as starch, glucose and oligosaccharides increases interest in its potential importance.[41]

PLA can be prepared in two different ways:

(i) Direct polycondensation of lactic acid.
(ii) ROP of lactide (LA).

Using the polycondensation approach it is difficult to prepare PLA with high molecular weight, and so the polymer does not have wide practical use because of its poor mechanical properties.

The most important and general way to prepare high molecular weight PLA is through ROP of a lactide which contains two chiral centers (Scheme 4.3).[42] The stereochemistry of the polymer formed determines its physical and mechanical properties, as well as its rate of degradation, and it is therefore important to ensure stereocontrol of the polymerization process. Apart from seterochemical control, the process is usually very long and requires an inert atmosphere.

Jing *et al.* investigated the ROP of D,L-lactide (DLLA) in a domestic microwave oven under atmosphere.[43] They presented a reaction condition

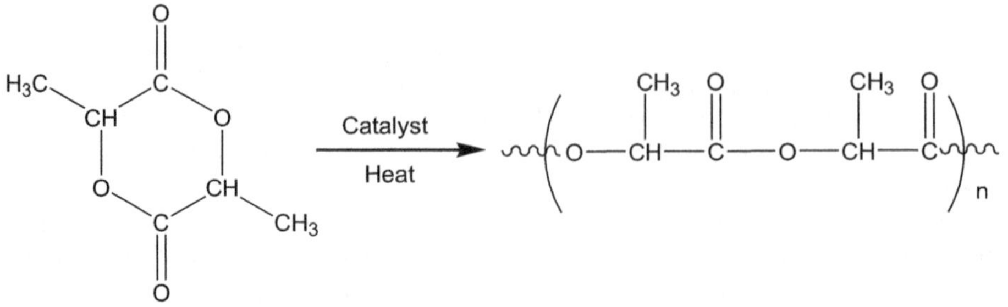

Scheme 4.3 Schematic representation of polylactide synthesis.[42]

dependent elaborate study on stannous octoate Sn(Oct)$_2$ catalyzed, ROP of DLLA. The reaction mixture was irradiated in a Gland domestic microwave oven (GuangDong, China, 2450 MHz and 800 W) at 450 W. After trying various reaction media like SiC, activated carbon and Al$_2$O$_3$, SiC was considered the most appropriate. The main advantages of SiC in the microwave field are that it can absorb microwaves effectively under low temperatures and keep them relatively stable.[44] On studying the effect of catalyst concentration it was found that the reaction rate increased on increasing the concentration, but not indefinitely. After a certain limit the reaction rate starts to decrease. This decrease was attributed to the fact that Sn(Oct)$_2$ catalyzes not only the polymerization of LA but also the decomposition of polymer.

Poly-D,L-lactide (PDLLA) with a viscosity average molecular weight 2.0×10^5 and a yield over 85% was obtained in a very short irradiation time (30 min). It was inferred that microwave-irradiated ROP of DLLA dismissed vacuum because of the improved selectivity and accelerated reaction rate. Comparing the temperature under microwave irradiation with that in conventional heating, it was considered that the enhanced polymerization rate originated not only from thermal effects but also from microwave effects.[43]

Frediani *et al.*[44] studied *cone*-25,27-dipropyloxy-26,28-dioxo-calix[4]arene titanium (IV) dichloride (Figure 4.3) in the ROP of DLLA. Calix[4]arenes are macrocyclic molecules made of four phenol units linked *via* methylene bridges connected to the *ortho* positions of the phenol rings. The presence of four oxygen atoms at the lower rim of these conical molecules provides a valuable platform for the synthesis of poly(phenoxy) metal complexes.

The most striking feature of the polymers formed using this catalyst is their partial isotactic-stereobloc microstructure, which is likely controlled by a chain end mechanism.

The use of microwave makes the synthesis of PDLLA by bulk polymerization process using tin(II) 2-ethylhexanoate as initiator much faster and energy saving, with a reaction time of about 30 minutes at 100 °C. The same synthesis by conventional process takes around 30 hours at 120 °C.[40] Application of sodium and potassium complexes in ROP of L-lactide was reported recently.[45] With these breakthroughs in the non-toxic and green catalysis, the ROP of lactides can find much better applications.

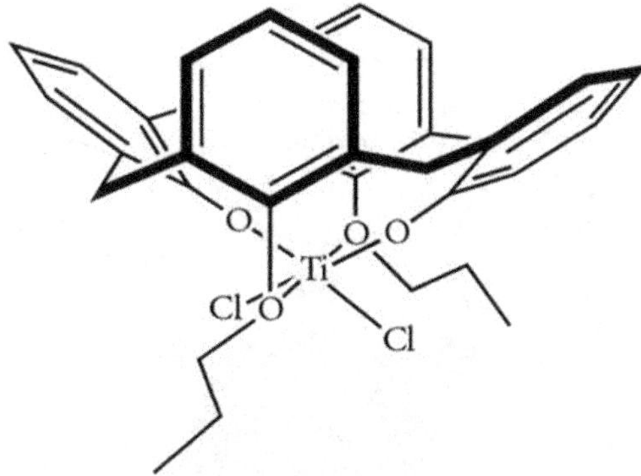

Figure 4.3 Structure of *cone*-25,27-dipropyloxy-26,28-dioxo-calix[4]arene titanium (iv) dichloride.

4.3 Enzyme Catalyzed Ring-opening Polymerization

As discussed in the previous sections, biodegradable polymers (mainly polylactones and polylactides) find several applications in the field of biomedicines, however, the toxicity of catalysts poses a great challenge to their widespread therapeutic use. Organometallic compounds are commonly used catalysts for ROP of cyclic lactones but their complete removal is rarely achieved. The residual catalyst causes latent toxicity for cell and organ.[46] *In vitro* enzymatic synthetic pathway has been presented as a panacea to this woe. The procedure consists of an *in vitro* polymerization *via* non-biosynthetic pathways catalyzed by an isolated enzyme. The process is nontoxic, requires mild reaction conditions, is recyclable and can be regioselectively synthesized by a one-pot lipase-catalyzed mechanism.

In order to fully comprehend the enzymatic mechanism, it is essential to know the difference between *in vivo* and *in vitro* reactions. *In vivo* reactions proceed under very mild, life-like conditions *i.e.*, under water as a media, neutral pH and low temperatures. The enzymatic reactions performed in laboratory conditions are generally carried out in bulk using non-aqueous, organic solvents and high temperatures. The difference in the reaction conditions is the major cause of lesser efficiency of *in vitro* reactions.[47] The organic solvents used deactivate the enzymes apart from causing solubility problems. Another reason for lower *in vitro* activity of enzymes can be attributed to their possible denaturation by organic solvents and also the lesser availability of active sites. We can thus infer that with the optimization of water content, the pH of the reactive media and choice of solvent and temperature, we can expect a major breakthrough in enzyme catalyzed ROP and other reactions.

The major advantages of enzyme catalyzed polymerization over chemical polymerizations[48] can be summed up in the following schematic presentation (Figure 4.4).

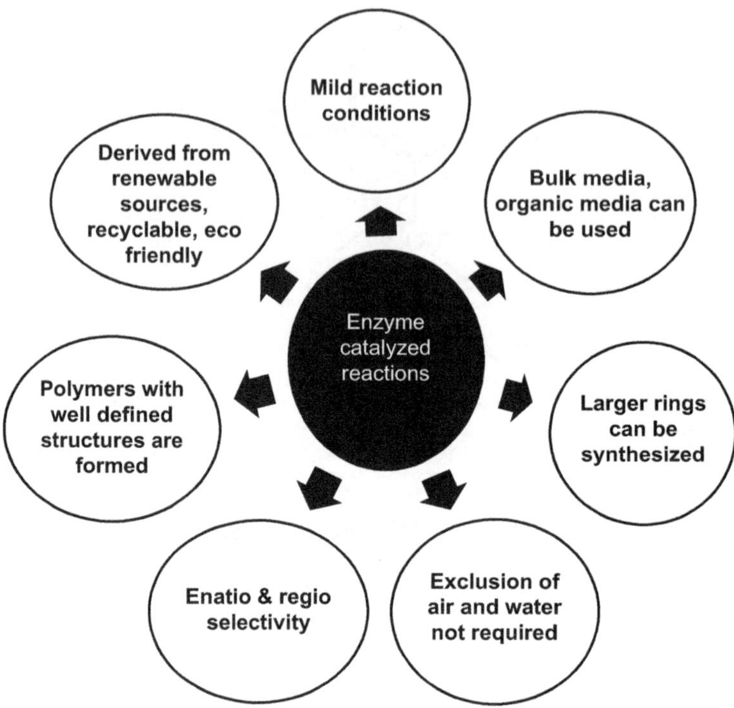

Figure 4.4 Advantages of microwave-assisted enzyme catalyzed reactions.

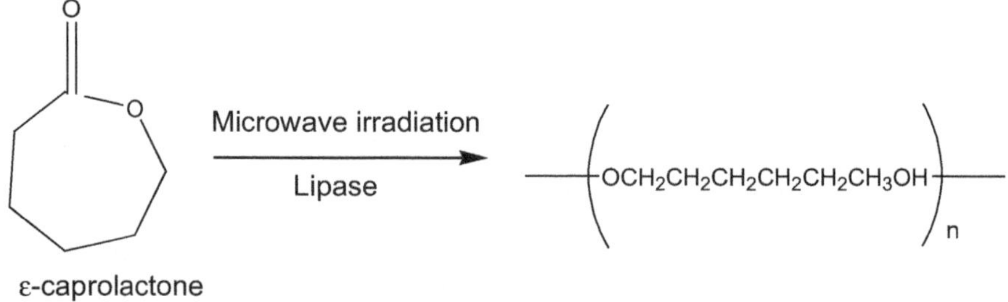

Scheme 4.4 Enzyme catalyzed microwave-assisted ring-opening polymerization of ε-caprolactone.[49]

Despite their great potential, enzyme catalyzed microwave reactions remain largely unexplored. Uyama and Kobayashi[49] reported in 1993 the enzyme-catalyzed polymerization of δ-valerolactone and ε-caprolactone by using lipase enzyme, *P. flourescens*, *Candida cylindracea* and porcine pancreatic lipase (PPL) (Scheme 4.4). After carrying out bulk polymerization for 10 days, Lipase *P. flourescens* gave the highest monomer conversion (92%) to obtain poly(3-CL) of $M_N = 7700$, with carboxyl and hydroxyl as end-groups. Several groups henceforth tried several combinations of solvents, reaction temperature and initiators for polymerization of caprolactum. Knani *et al.*[50]

carried out the polymerization of ε-CL in *n*-hexane solution in the temperature range of 25–40 °C using crude porcine pancreatic lipase (PPL) with methanol as preferred nucleophile. Gross and co-workers[51] investigated the solution polymerization of ε-CL using dioxane, toluene and heptanes as solvent, PPL as catalyst and butanol as initiating species.

Unsubstituted lactones with a ring-size from 4 to 17 have since been polymerized by using lipases from *Aspergillus niger* (lipase A), *Candida antartica* (lipase B, Novozyme-435), *Candida rugosa* (lipase AYS), *Mujor javancius* (lipase M), *Mucor meihei* (lipozyme), porcine pancreatic lipase (PPL), *Pseudomonas aeruginosa* (lipase PA), *Pseudomonas cepacia* (lipase PS), *Pseudomonas fluorescens* (lipase PF), *Pseudomonas* sp. lipase (PSL), *Rhizopus delemar* (lipase RD) and *Rhizopus japonicus* (lipase RJ), *etc.* The enzyme's ability to support the development of immobilized lipase catalyst for the efficient production of polyesters has been investigated.[52] The preferred lipase system used is generally a physically immobilized form of *Candida antartica* known as Novozyme-435. Porous ceramic (silica) as well as polymeric supports have been evaluated. Porous polypropylene was found to be good support for *Candida antartica* in the polymerization of 15-pentadecanolide.[53]

The influence of MW irradiation on the enzymatic ROP of ε-caprolactone using Novozym 435 was studied by Kerep *et al.*[54,55] This study exhibited the formation of a higher amount of polyester with a terminal SH moiety and the change in chemoselectivity. Feng *et al.*[56] polymerized 26-membered macrocyclic carbonate, cyclobis(decamethylene carbonate) [(DMC)$_2$] by ROP using Novozym-435 in toluene. High molecular weight (Mn) of 5.4×10^4 and yield of 99% was obtained at ultra-low enzyme/substrate (E/S) weight ratio of 1/200. The reaction exhibited pseudo-living characteristics and followed first order reaction kinetics, as shown in Scheme 4.5 below.

Despite the great promise they hold, there are several problems associated with *in vitro* biocatalytic polymerization which hinder the large-scale employment of enzymes in polymer production by the industry. They can be listed as follows:

(a) High cost of enzyme.
(b) Large quantity of enzyme required.
(c) Long duration of reactions.
(d) Formation of relatively low molecular weight products.

We can thus say that the need of the hour is to address these challenging problems before the industrialization of the process. Some of the remedies presented for the above include covalent modification of enzymes to provide highly active and organic soluble enzymes. This may help in increasing the efficiency of the enzymes. Immobilization and protein engineering are known to improve the biocatalyst's efficiency and stability. Addition of inorganic salts has been shown to improve the enzyme activity. The ultimate goals are to improve the enzyme's activity, stability and selectivity. Future works in these directions will certainly provide the polymer chemist with a

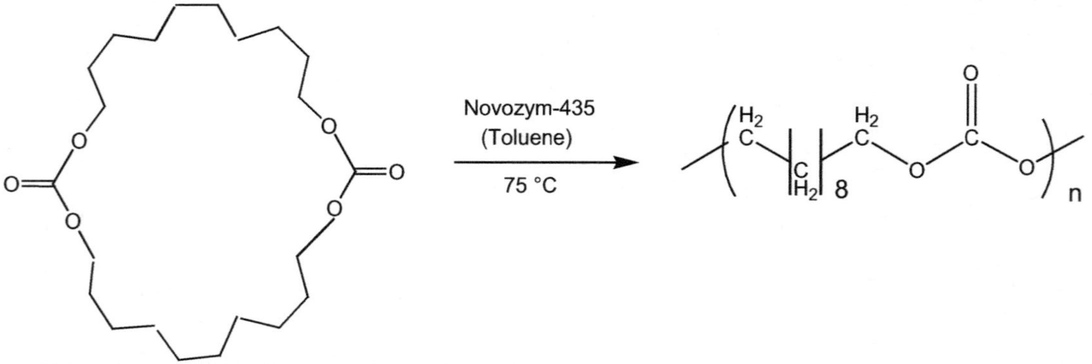

26-membered cyclobis (decamethylene carbonate)

Novozym-435
(Toluene)
75 °C

Scheme 4.5 Enzyme catalyzed ring-opening polymerization of cyclobis (decamethylene carbonate).[56]

biocatalyst which can function as efficiently in polymer synthesis as *in vivo* enzymes are able to do.[48]

4.4 Cationic/Anionic Ring-opening Polymerizations

Living polymerization is a form of addition polymerization where the ability of a growing polymer chain to terminate has been removed. In living polymerization reactions chain termination and chain transfer reactions are absent and the rate of chain initiation is also much larger than the rate of chain propagation. The result is that the polymer chains grow at a more constant rate than that seen in traditional chain polymerization and their lengths remain very similar (*i.e.* they have a very low polydispersity index). Living polymerization is a popular method for synthesizing block co-polymers since the polymer can be synthesized in stages, each stage containing a different monomer. Additional advantages are predetermined molar mass, specific functionalities, customized architecture and control over end-groups.[57] According to the reaction mechanism, living polymerization can be classified into two categories:

(1) living radical polymerization;
(2) living cationic ring-opening polymerization.

Living radical polymerization (LROP) has been studied intensively under microwave irradiation,[58] showing advantages like increased reaction rates and improved polymer properties. MW-assisted LROP using 2-oxazolines as monomers was reported recently.[59] Poly(2-oxazolines) have attracted much interest because of the biological activity displayed by many molecules possessing this core structure, which find applications in micelle catalysis, drug delivery and hydrogels. The LROP of 2-oxazolines is a slow process with a reaction time ranging from hours to days; thus, a reduced reaction time is a challenge to industrial applications of the process. MW-assisted LROPs of 2-methyl-2-oxazoline (MeOx), 2-ethyl-2-oxazoline (EtOx), 2-nonyl-2-oxazoline (NonOx), 2-phenyl-2-oxazoline (PhOx) and soy based 2-oxazoline were investigated and the polymerization time had been reduced to only a few minutes (Scheme 4. 6).[60]

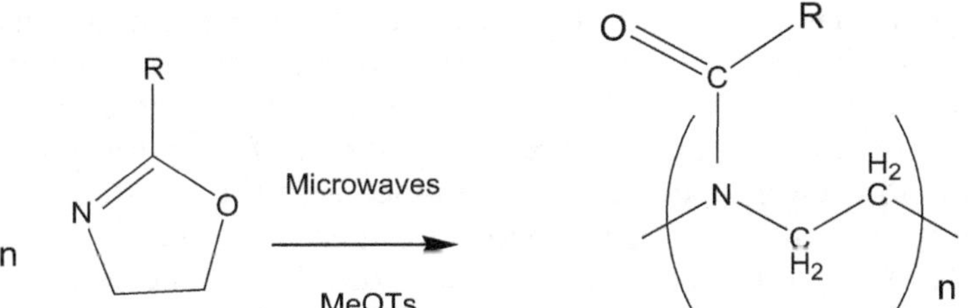

Scheme 4.6   Microwave-assisted   cationic   ring-opening   polymerization   of 2-oxazolines.[60]

Schubert and co-workers were the first to report methyl tosylate (TsOMe) initiated MW-assisted LROP of EtOx.[61] They compared the kinetics of the reaction under both conventional and microwave conditions. The reaction was carried out in an Emrys Liberator single-mode microwave oven. The six-hour conventional polymerization was performed at 80 °C in acetonitrile. In contrast, the MW irradiated reaction was quick, homogeneous and required a minimum amount of solvent. The advantage of using a minimal amount of solvent includes the attainment of high temperatures (up to 200 °C), independent of the boiling point of the solvent. The living nature of the MW-assisted polymerization was observed in the temperature range between 80 and 180 °C, and polymers were obtained with a narrow M_W distribution (PDI, 1.2). The kinetics showed that the reaction rate increased with temperature and the reaction at 190 °C under microwave irradiation was completed in just one minute (almost 350 times faster) than that by conventional heating at 80 °C. It would be worth mentioning that activation energy, however, remained unaltered. The increase in reaction rate was attributed to thermal microwave effects.

Sinnwell *et al.*[62] and Schubert *et al.*[63] reported contradictory results about the thermal and nonthermal microwave effects in the MW-assisted LROP of PhOx. The first group used four systems to study the LROP of PhOx: (1) a conventional heating closed system (CCS); (2) a conventional heating open system (COS); (3) a MW irradiation closed system (MCS); and (4) a MW irradiation open system (MOS). In the closed system (CS), the monomer and TsOMe were dissolved in acetonitrile and the reaction temperature was approximately 125 °C. In the open system (OS), butyronitrile was used as solvent and the temperature of the reaction medium was approximately 123 °C. In all four systems, the linear plots of $\ln([M_0]/[M_t])$ (where $[M_0]$ is the initial monomer concentration and $[M_t]$ is the concentration of monomer after time, t, against time showed that the reaction rate follows the first-order kinetics. MW-assisted LROP showed a great enhancement in reaction rates compared with the thermal LROP. It was observed that the microwave reactions in both open and closed systems showed nearly the same enhancement rate as compared to corresponding thermal systems. The similar enhancement of the reaction rate was attributed to the strong MW effect that took place mainly in the active growing polymer (Scheme 4.7a), which was composed of a tosylate counterion of the active oxazolinium end group (Scheme 4.7b). The authors further measured the increase in temperature of the reaction mixtures by varying monomer-to-initiator ratios corresponding to a variable ratio of ionic species. The temperature increased with the amount of the initiator, and the final temperature of the pure solvent (80 °C) is considerably lower, which is in accordance with the low dielectric loss of acetonitrile. This interesting behavior indicates that there is a strong MW effect that mainly takes place at the ionic oxazolinium species.

Schubert's group attributed the acceleration of the LROP of PhOx under microwave irradiation only to the temperature effects.[63] They studied LROP of PhOx in three systems: (1) MCS (in butyronitrile); (2) automated CCS

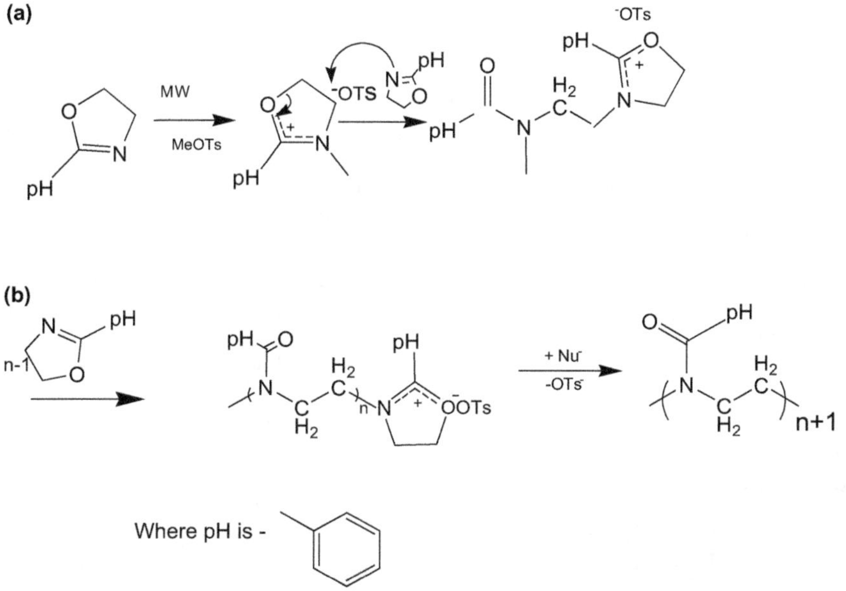

(a)

(b)

Where pH is -

Scheme 4.7 Cationic ring-opening polymerization of 2-phenyl-2-oxazoline using (a) methyl tosylate and (b) active oxazoline polymer.[62,63]

(under pressure, in acetonitrile); and (3) automated COS (under reflux, in butyronitrile). Both pressure polymerizations in acetonitrile and reflux polymerization in butyronitrile revealed similar polymerization rates under microwave irradiation and conventional heating. The first order kinetic plots of the three systems and good correspondence in the gel permeation chromatography (GPC) traces of the polymerization mixtures demonstrated the absence of the non-thermal microwave effects.

Based on the above results, a detailed study on the kinetics and the living nature of the LROP were performed on a series of linear 2-oxazolines, including MeO_x, EtO_x, $NonO_x$ and PhO_x in acetonitrile at high temperatures of up to 200 °C under microwave irradiation.[64] The reaction rates of MLROP were enhanced significantly by factors of up to 400 times that of the conventional LROP. The first-order kinetics of the monomer consumption and the livingness of the polymerization were maintained in MLROP at temperatures ranging from 80–200 °C. The activation energies for the MLROP of the four 2-oxazolines were found to fall in the range of the values obtained with conventional heating.

Soy based 2-oxazoline (SoyOx) monomer[65] contains unsaturated groups that can provide cross-links for the resulting polymers. TsOMe initiated LROP of the SoyOx also was carried out under microwave irradiation in bulk or in acetonitrile. The polymerization exhibited living characteristics and was completed in just 15 min at 140 °C in acetonitrile. These investigations on MW-assisted LROP of 2-oxazolines conducted by Schubert and co-workers

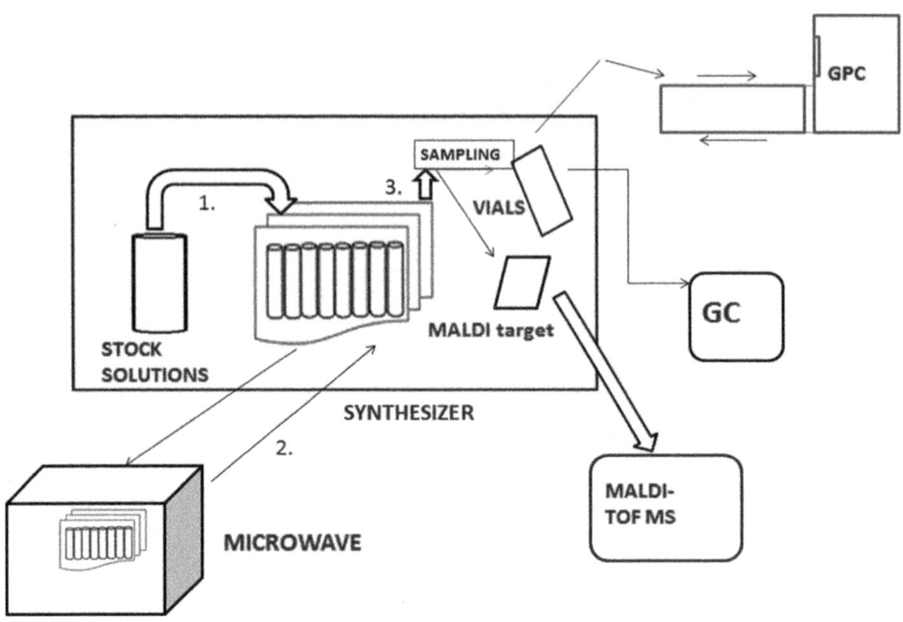

Figure 4.5 Microwave high-throughput work flow.[66]

further affirmed that only thermal effects were responsible for the tremendous increase in the 2-oxazoline polymerization rate.

Hoogenboom *et al.*[66] reported a microwave-assisted high throughput method (where different parameters can be screened simultaneously or in a fast serial mode and the results can be compared easily) for LROP of 2-oxazolines. The system included a single-mode microwave synthesizer and high throughput workflow (high-throughput screening with peripheral characterization equipment) (Figure 4.5). MW-assisted LROP of NonO$_x$ at 140 °C was carried out in solvent, and kinetic investigations in dichloromethane revealed a living mechanism. Polymerizations under MW irradiation proceeded significantly faster than conventionally heated polymerizations (at ambient pressure) because they combined the advantages of both MW-assisted polymer synthesis and high-throughput polymer synthesis.

4.5 Ring-opening Copolymerization

A heteropolymer or copolymer is a polymer derived from two (or more) monomeric species, as opposed to a homopolymer, where only one monomer is used.[67] Copolymerization refers to methods used to chemically synthesize a copolymer. Copolymerization is the simultaneous polymerization of two or more types of monomers and it is a useful method to obtain polymers with adjustable properties. In ring-opening copolymerizations (ROCP), microwaves have been introduced to prepare random copolymers and block copolymers. The following section discusses two kinds of co-polymers (copolymers based on lactones and 2-oxazolines, respectively) synthesized by MW-assisted ring-opening copolymerization (MROCP).

4.5.1 Ring-opening Copolymerization of Lactones/Lactides

The MW-assisted copolymerizations of ε-caprolactone and ε-caprolactame were reported by Fang *et al.*[68] They compared microwave-assisted anionic copolymerization with the conventional heating method using lithium tri-*tert*-butoxyaluminohydride as a catalyst. They studied the effects of catalyst concentration, reaction temperature and time on the copolymer composition and thermal properties. A variable frequency microwave furnace model LT 502 Xb (Lambda Technologies, Inc.) with a frequency range from 2.4 to 7.0 GHz was used to synthesize poly(caprolactam-*co*-caprolactone). The copolymerization temperature was measured and controlled *via* a grounded Omega K-type thermocouple inserted in the sample. The thermocouple was initially calibrated against a Luxtron optic temperature probe. The reaction is outlined in Scheme 4.8.

Dielectric properties revealed that both ε-caprolactam and ε-caprolactone gave an effective absorption of electromagnetic energy over the microwave range in the liquid state. MW synthetic results showed amide composition increased with the increased reaction temperature, time and catalyst level, while molecular weights of copolymers decreased with the increased catalyst levels. Compared with the corresponding thermal products, microwave-synthesized copolymers gave higher yield, higher amide composition, higher glass transition temperature and equivalent molecular weight.

Chitosan is known to be biocompatible, enzymatically biodegradable, nontoxic and bioactive in animal tissues. However, poor mechanical and processing properties and lesser solubility in organic solvents as well as aqueous solutions at the physiological pH limit its applications. Grafting copolymerization of chitosan with other monomers like CL, LA, *etc.* lead to drastic improvements in its inherent limitations and hence applicability.

Liu and co-workers[69] reported the microwave-assisted graft ROCP of CL onto chitosan. The reaction was carried out by a protection-graft-deprotection procedure, with phthaloylchitosan as the precursor, and $Sn(Oct)_2$ as the catalyst (Scheme 4.9).

After being irradiated at 450 W for 15 min, a high grafting percentage of 232% was achieved. The graft of CL onto chitosan was greatly accelerated and enhanced under microwave irradiation, and the graft copolymerization was greatly improved at the higher power outputs of 450 and 600 W in the studying range. After deprotection, the phthaloyl group was removed and the amino group was regenerated. Thus, the chitosan-*g*-polycaprolactone co-polymer was an amphoteric hybrid with both a large amount of free amino groups and hydrophobic polycaprolactone side chains.

The MW-assisted direct polycondensation of LA to synthesize PLA has been shown to be highly accelerated as compared to conventional heating.[70]

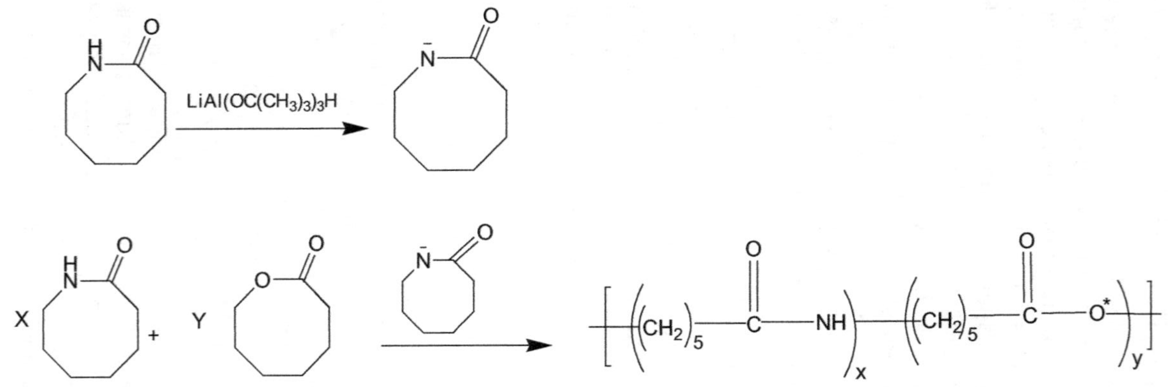
LiAl(OC(CH3)3)3H
X
Y
+
NH
x
y

Scheme 4.8 Stepwise anionic ring-opening copolymerization mechanism of ε-caprolactam and ε-caprolactone.[68]

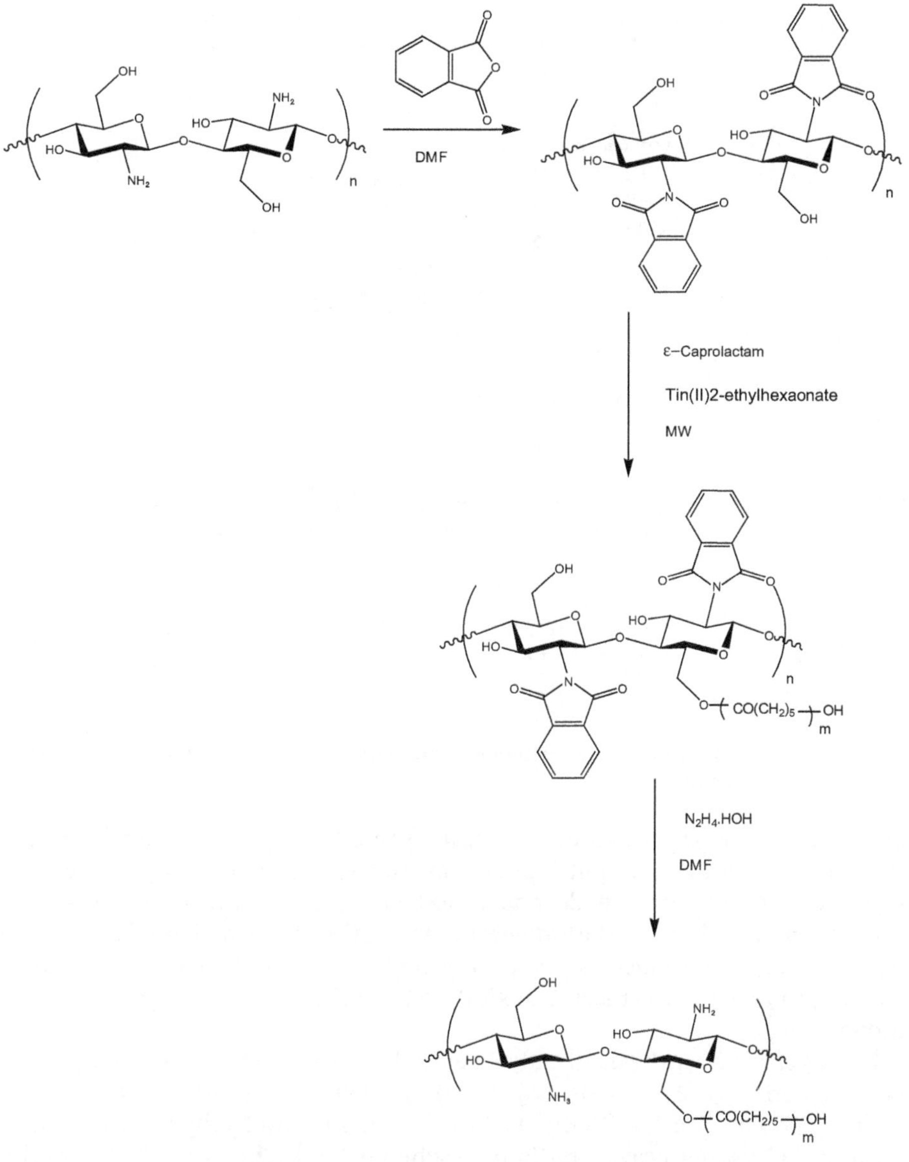

Scheme 4.9   Graft copolymerization of chitosan with ε-caprolactam. The protection and deprotection approach ensured the polymerization at/from hydroxyl functional group.[69]

This single-step direct polycondensation method generated PLA with M_W 16 000 using $SnCl_2/p$-TsOH binary catalyst under reduced pressure of around 30 mmHg. The scheme of reaction is shown in Scheme 4.10 below.

The polymer obtained is colorless and has enough strength to make pellets and films, which can be applied as a raw material of biodegradable

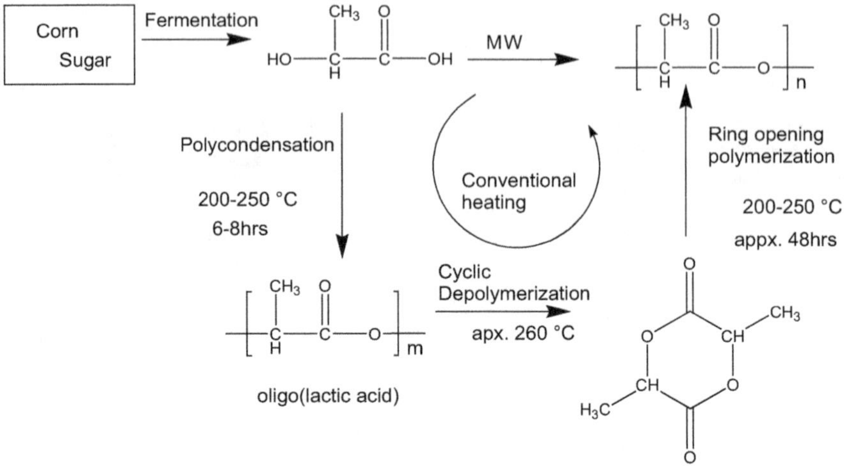

Scheme 4.10 Schematic presentation of poly lactic acid synthesis, comparison drawn between microwave and conventional process.[70]

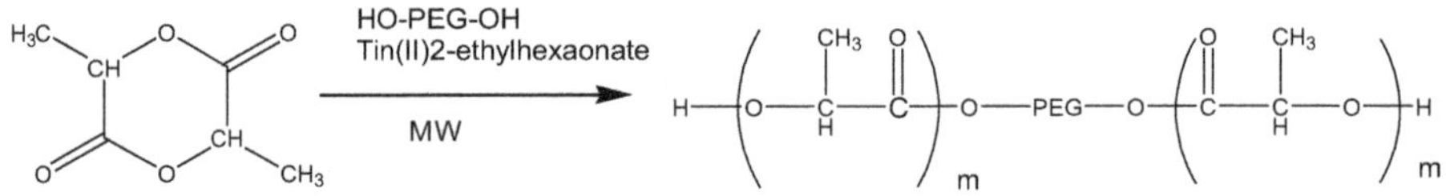

Scheme 4.11 PLLA-PEG-PLLA triblock copolymer synthesis under microwave irradiation.[72]

plastics. This polymer can also be transformed easily to high-performance PLA with a molecular weight higher than 100 000 by melt polycondensation, which can then be used widely as a bio-based plastic in various fields.

Copolymerization of ε-caprolactone with ethylene glycol produced well-defined poly(ε-caprolactone)-*block* poly(ethylene glycol)-*block*-poly(ε-caprolactone) tri-*block* copolymers that showed the ability to release ibuprofen as a model drug.[71]

Zhang *et al.*[72] synthesized poly(L-lactide)-*block*-poly(ethylene glycol)-*block*-poly(L-lactide) (PLLA-PEG-PLLA) by poly(ethylene glycol) initiated ring-opening polymerization (ROP) of L-lactide using tin (II) 2-ethylhexanoate as a catalyst under microwave irradiation (Scheme 4.11). The introduction of the hydrophilic PEG segment into the hydrophobic PLLA chains would lead to copolymers with controlled hydrophilicity and degradability; both of these properties could prove to be highly useful in biomedicines as well as in the packaging industry.

The 92.4% yielding reaction gave polymer having a number-average molar mass of 28 230 g mol^{-1} and in 3 min at 100 °C. The L-lactide/poly(ethylene glycol) feed ratios had a strong influence on the polymerizations. A higher feed ratio can result in copolymers with higher molar masses and lower L-lactide conversions. However, prolonged microwave irradiation did not have any significant effect on the aforementioned aspects. The thermal property investigation suggested that the glass transition temperature and the melting temperature of the PLLA-PEG-PLLA tri*block* copolymers increased with the PLLA segment length in the copolymer.

Grafting copolymerization of D,L-lactide onto chitosan leads to better physico-chemical properties than the homopolymers. Luo *et al.*[73] synthesized brush-like chitosan-*g*-poly(D,L-lactide) copolymers by bulk ring-opening polymerization of D,L-lactide(D,L-LA) using chitosan as macro-initiator under microwave irradiation (Scheme 4.12).

With the help of FT IR, 1H NMR, EA, XRD and TGA characterizations, it was inferred that the crystalline nature and thermal stability of chitosan decreased after grafting, which is expected to improve its solubility and processing property.

Amphiphilic guar gum grafted with poly(ε-caprolactone) (GG-*g*-PCL) was fabricated as a drug-delivery carrier using microwave irradiation. The GG-*g*-PCL copolymer is capable of self-assembling into nano-sized spherical micelles in aqueous solution. A series of biodegradable poly(vinyl alcohol)-graft-poly(ε-caprolactone) comb-like polyester were prepared in bulk. The introduction of hydrophilic backbone resulted in the decrease in both the melting point and crystallization property of the PVA-*g*-PCL copolymers comparing with linear PCL.[5,74] A thermoformable starch-*graft*-poly-caprolactone biocomposite was prepared by initiating ring-opening poly-merization of caprolactone monomer onto starch under microwave irradiation. In this case, the thermoplastic and hydrophobic modification of starch could be realized by one-pot grafting of PCL, where the grafted PCL chains acted as the ''plasticizing'' tails of thermoforming and as the hydrophobic species of water-resistance.[75]

4.5.2 Ring-opening Copolymerization of 2-oxazolines

Copolymerization of 2-oxazolines combine the advantages of the living polymerization, such as the good control of composition and structure of the resulting polymer, with the possibility to synthesize a polymer scaffold that can be functionalized with multiple homing devices for improved target-ing,[76] the control of the solubility, biodistribution, pharmacokinetics and immunoresponse of the entire construct by adjusting the polymer com-position. For example, amphiphilic block co-poly(2-oxazoline)s can be ob-tained by a combination of a hydrophilic *block* poly(2-ethyl-2-oxazoline) and a hydrophobic *block* poly(2-nonyl-2-oxazoline). Schubert and co-worker[77] expanded the LROHP of 2-oxazolines to the LROCP in preparation of block co-poly(2-oxazolines), using four monomers: MeOx, EtOx, NonOx and PhOx (Scheme 4.13).

This LROCP was initiated by TsOMe and performed in acetonitrile at 140 °C in a single-mode microwave reactor. A total number of 100 (50 + 50) monomer units were incorporated into the polymer chains, resulting in a library of four chain-extended homo- and 12 di*block* *co*-poly(2-oxazoline)s

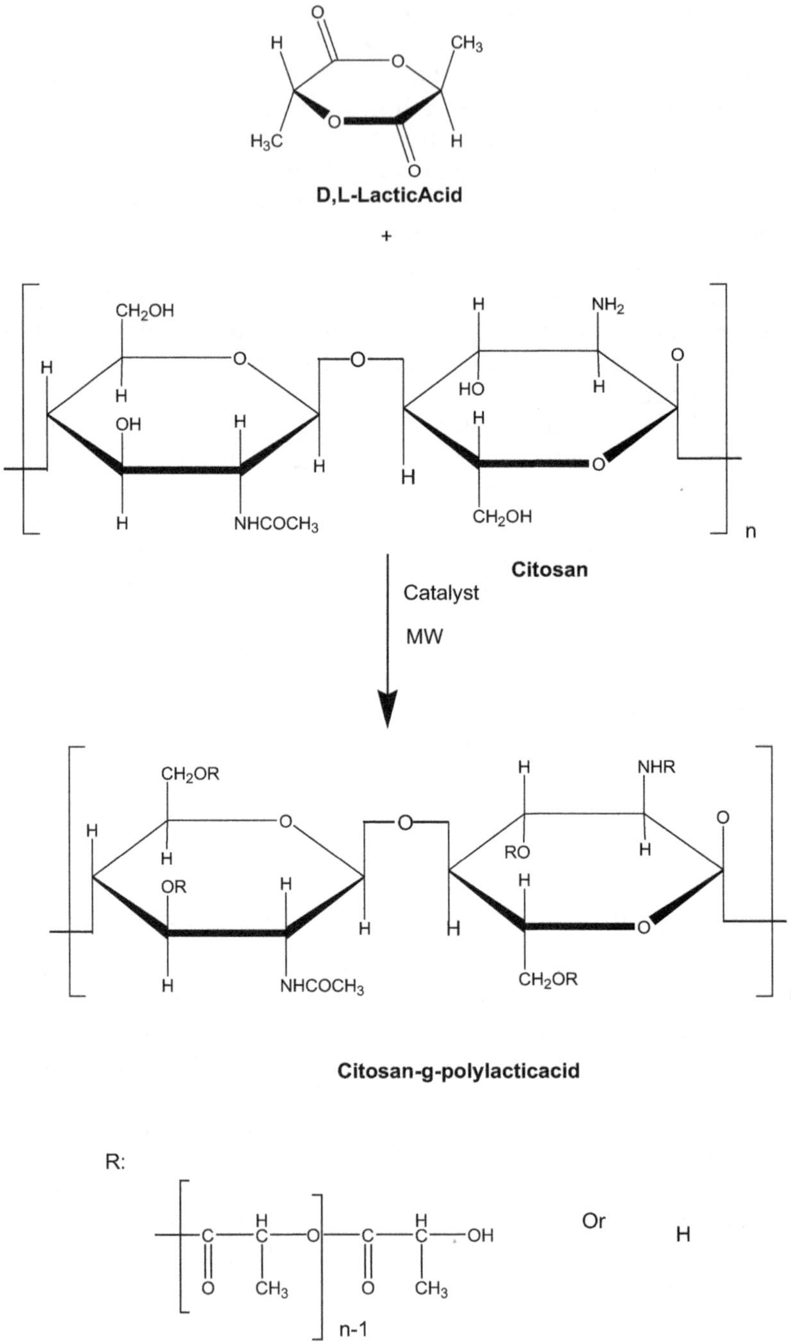

Scheme 4.12   Synthesis of chitosan-grafted-poly(D,L,-Lactide) copolymer.[73]

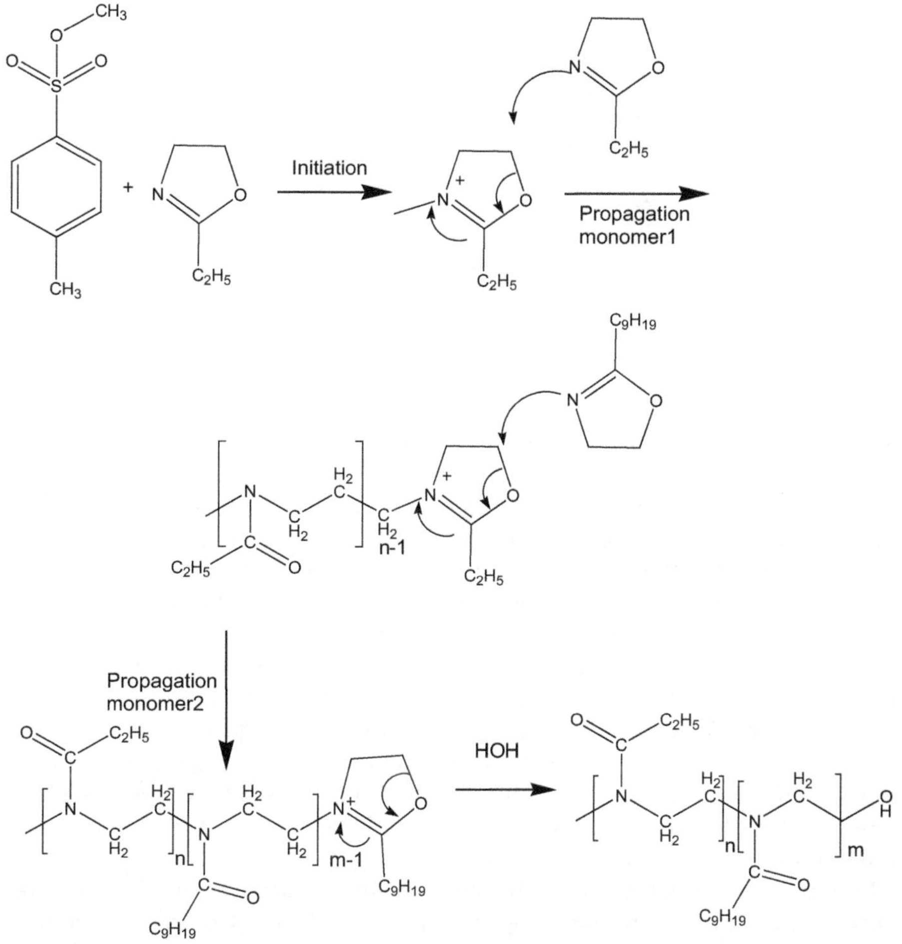

Scheme 4.13 Mechanism for the stepwise synthesis of et50-non50 *block* copolymer from the cationic ring-opening polymerization of 2-ethyl- and 2-nonyl-2-oxazoline, initiated by methyl tosylate.[76]

with narrow distributions (PDI, 1.30). The polymerization kinetics of LROCP of 2-oxazolines in acetonitrile was investigated under different pressures. In addition, a series of *block* copolymers was synthesized in an automated parallel synthesis robot by this pressure polymerization method. These studies indicate that the microwave copolymerization of 2-oxazolines was an efficient method for preparing copolymers in batch.

The possibilities of up-scaling the cationic ROP of 2-ethyl-2-oxazoline were investigated by both a batch-type and a continuous-flow approach (Figure 4.6).[76,78]

Further improvements were found by using ionic liquids (IL) instead of common organic solvents as solvents for the MW-assisted cationic ROP of 2-ethyl-2-oxazoline.[79] The living cationic ring-opening polymerization

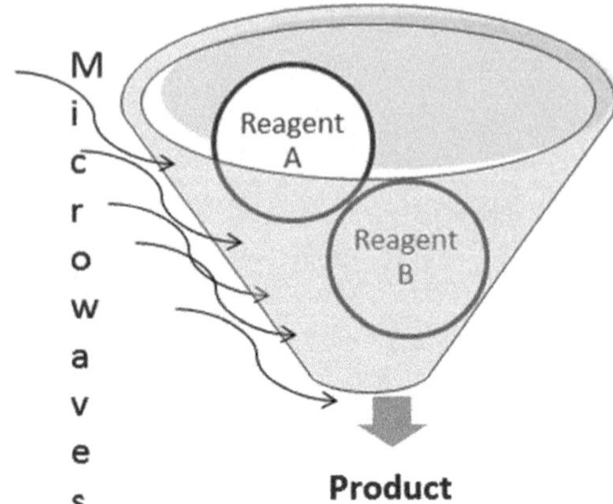

Figure 4.6   Pictorial depiction of continuous flow approach for synthesis.

of 2-ethyl-2-oxazoline performed in an ionic liquid under microwave irradiation showed an enhanced polymerization rate. It is important to note that ionic liquid was efficiently recovered and reused in new reaction cycles, completely avoiding the use of organic volatile compounds. It is known that ILs containing $[PF_6]^-$ or $[BF_4]^-$ anions may hydrolyze under acidic conditions and release hydrogen fluoride.[80] However, this could be prevented by closely monitoring the pH throughout the reaction. Besides, in order to avoid completely the formation of strong acids in the entire process, more stable hydrophobic ILs may be tested for the same purpose, for example, 1-hexyl-3-methylimidazolium tris(pentafluoroethyl)trifluorophosphate). We can thus conclude that the use of ionic liquids in polymerization is a budding field and can open up a new window for green synthesis.

In microwave heated ROP, the convection/conduction heating through conventional mode is replaced by volumetric heating. Most of the scientific reports conclude that volumetric heating is most probably responsible for the benefits obtained in MW heated reactions. However, many reports highlighted negative effects on rate and yield of reaction. The role of solvents employed for the reaction also seems important. However, it can definitely be concluded that the ROP conducted under MW irradiation gave impressive results that encourage the scientific community to further explore the field to find even better results.

References

1. A. D. Jenkins, P. Kratochvíl, R. F. T. Stepto and U. W. Suter, Glossary of basic terms in polymer science (IUPAC Recommendations 1996), *Pure Appl. Chem.*, 1996, **68**(12), 2287.
2. C. Zhang, L. Liao and S. Gong, *Green Chem.*, 2007, **9**, 303.
3. G. Odian, *Principles of Polymerization*, John Wiley & Sons, Inc., 4th edn, 2004, ch. 7, p. 544.

4. K. Kempe, C. R. Becer and U. S. Schubert, *Macromolecules*, 2011, **44**, 5825.
5. A. Mishra and R. Dubey, *Green Chemistry for Environmental Remediation*, ed. R. Sanghi and V. Singh, John Wiley & Sons, Inc., Hoboken, NJ, USA, 1st edn, 2012, ch. 13, p. 404.
6. G. Sivalingam, N. Agarwal and G. Madras, *J. Appl. Polym. Sci.*, 2004, **91**, 1450.
7. A. C. Albertsson and I. K. Varma, *Biomacromolecules*, 2003, **4**, 1466.
8. X. Fang, R. Hutcheon and D. A. Scola, *J. Polym. Sci., Part A: Polym. Chem.*, 2000, **38**, 1379.
9. K. Aoi and M. Okada, *Prog. Polym. Sci.*, 1996, **21**, 151.
10. Ph. Dubois, C. Jacobs, R. Jerome and Ph. Teyssie, *Macromolecules*, 1978, **11**, 68.
11. H. R. Kricheldorf, I. K. Saunders and C. Boettcher, *Polymer*, 1995, **36**, 1253.
12. A. Kowalski, A. Duda and S. Penczek, *Macromol. Rapid Commun.*, 1998, **19**, 567.
13. S. Sosnowski, S. Slomkowski, S. Penczek and L. Reibel, *Macromol. Chem.*, 1983, **84**, 2159.
14. F. Stassin, O. Halleux and R. Jerome, *Macromolecules*, 2001, **34**, 775.
15. P. Albert, H. Warth, R. Mulhaupt and R. Janda, *Macromol. Chem. Phys.*, 1996, **197**, 1633.
16. X. M. Fang, S. J. Huang and D. A. Scola, *Polym. Mater. Sci. Eng.*, 1998, **79**, 518.
17. Q. Xu, C. Zhang, S. Cai, P. Zhu and L. Liu, *J. Ind. Eng. Chem.*, 2010, **16**, 872.
18. L. Q. Liao, L. J. Liu, F. He, R. X. Zhuo and K. Wan, *J. Polym. Sci., Part A: Polym. Chem.*, 2002, **40**, 1749.
19. S. Sinnwell, A. M. Schmidt and H. Ritter, *J. Macromol. Sci., Part A: Pure Appl. Chem.*, 2006, **43**, 469–476.
20. I. Barakat, Ph. Dubois, R. Jerome and Ph. Teyssie, *Macromolecules*, 1991, **24**, 6542.
21. J. Okuda and I. L. Rushkin, *Macromolecules*, 1993, **26**, 5530.
22. T. Ouhadi, C. Stevens and Ph. Teyssie, *Mackromol. Chem. Suppl.*, 1975, **1**, 191.
23. M. Endo, T. Aida and S. Inoue, *Macromolecules*, 1987, **20**, 2982.
24. Y. Shen, Z. Shen, J. Shen, Y. Zhang and K. Yao, *Macromolecules*, 1996, **29**, 3441.
25. H. Yasuda and H. Tamai, *Prog. Polym. Sci.*, 1993, **18**, 1097.
26. A. Kowalski, A. Duda and S. Penczek, *Macromol. Rapid Commun.*, 1998, **19**, 567–572.
27. R. F. Storey and A. E. Taylor, *J. Macromol. Sci., Part A: Pure Appl. Chem.*, 1996, **33**, 77.
28. R. F. Storey and A. E. Taylor, *J. Macromol. Sci., Part A: Pure Appl. Chem.*, 1998, **35**, 723.
29. L. Q. Liao, L. J. Liu, C. Zhang, X. L. Wang and R. X. Zhuo, *Chin. Chem. Lett.*, 2001, **12**, 663.

30. I. Kreisersaunders and H. R. Kricheldorf, *Macromol. Chem. Phys.*, 1998, **199**, 1081.

31. G. A. Abraham, A. Gallardo, A. Lozano and J. S. Roman, *J. Polym. Sci., Part A: Polym. Chem.*, 2000, **38**, 1355.

32. Z. J. Yu, L. J. Liu and R. X. Zhuo, *J. Polym. Sci., Part A: Polym. Chem.*, 2003, **41**, 13.

33. Z. J. Yu and L. J. Liu, *Eur. Polym. J.*, 2004, **40**, 2213.

34. Y. Song, L. J. Weng, X. C. Liu and R. X. Zhuo, *J. Biomater. Sci., Polym. Ed.*, 2003, **414**, 241.

35. Y. Song, L. J. Liu, Z. J. Yu, X. C. Weng and R. X. Zhuo, *Polym. Prepr. (Am. Chem. Soc., Div. Polym. Chem.)*, 2003, **44**, 936.

36. Y. Song, L. J. Liu and R. X. Zhuo, *Chin. Chem. Lett.*, 2003, **14**, 32–34.

37. S. Aime, M. Botta, M. Fasano and E. Terreno, *Chem. Soc.Rev.*, 1998, **27**, 19.

38. D. Barbier-Baudry, C. H. Brachais, A. Cretu, A. Loupy and D. Stuerga, *Macromol. Rapid Commun.*, 2002, **23**, 200.

39. N. T. Nguyen, E. Greenhalgh, M. J. Kamaruddin, J. Elharfi, K. Carmichael, G. Dimitrakis, S. W. Kingman, J. P. Robinson and D. J. Irvine, *Tetrahedron*, 2014, **70**, 996.

40. S. Gong, L. Liao and C. Zhang, *Macromol. Chem. Phys.*, 2007, **208**, 1122.

41. N. Ljubisa, R. Ivan, A. Borivoj, N. Vesna, J. Jelena and S. Mihajlo, *Sensors*, 2010, **10**, 5063.

42. B. Koroskenyi and S. P. McCarthy, *J. Polym. Environ.*, 2002, **10**, 93.

43. S. Jing, W. Peng, Z. Tong and Z. Baoxiu, *J. Appl. Polym. Sci.*, 2006, **100**, 2244.

44. M. Frediani, D. Šemeril, D. Matt, L. Rosi, P. Frediani, F. Rizzolo and A. M. Papini, *Int. J. Polym. Sci*, 2010, **2010**, 1.

45. L. Chen, L. Jia, F. Cheng, L. Wang, C. Lin, J. Wu and N. Tang, *Inorg. Chem. Commun.*, 2011, **14**, 26.

46. J. Feng, H. F. Wang, X. Z. Zhang and R. X. Zhuo, *Eur. Polym. J.*, 2009, **45**, 523.

47. A. M. Klibanocv, *Trends Biotechnol.*, 1997, **15**, 97.

48. I. K. Varma, A. C. Albertsson, R. Rajkhowa and R. K. Srivastava, *Prog. Polym. Sci.*, 2005, **30**, 949.

49. H. Uyama and S. Kobayashi, *Chem. Lett.*, 1993, **22**, 1149.

50. D. Knani, A. L. Gutman and D. H. Kohn, *J. Polym. Sci., Part A: Polym. Chem.*, 1993, **31**, 1221.

51. R. T. MacDonald, S. Pulapura, Y. Svirkin, R. A. Gross, D. A. Kaplan, J. Akkara and g. Swift, *Macromolecules*, 1995, **28**, 73.

52. H. Uyama, M. Kuwabara, T. Tsujimoto and S. Kobayashi, *Polym. J.*, 2002, **34**, 970.

53. S. Kobayashi, S. Shoda and H. Uyama, *Adv. Polym. Sci.*, 1995, **121**, 1.

54. P. Kerep and H. Ritter, *Macromol. Rapid Commun.*, 2006, **27**, 707.

55. P. Kerep and H. Ritter, *Macromol. Rapid Commun.*, 2007, **28**, 759.

56. J. Feng, H. Wang, X. Zhang and R. Zhuo, *Eur. Polym. J.*, 2009, **45**, 523.

57. http://www.google.co.in/url?q = http://en.wikipedia.org/wiki/cationic_ polymerization.
58. D. D. Wisnoski, W. H. Leister, K. A. Strauss, Z. Zhao and C. W. Lindsley, *Tetrahedron Lett.*, 2003, **44**, 4321.
59. W. Xu, X. Zhu, Z. Cheng, G. Chen and J. Lu, *Eur. Polym. J.*, 2003, **39**, 1349.
60. C. Zhang, L. Liao and S. Gong, *Green Chem.*, 2007, **9**, 303.
61. F. Wiesbrock, R. Hoogenboom, C. H. Abeln and U. S. Schubert, *Macromol. Rapid Commun.*, 2004, **25**, 1895.
62. S. Sinnwell and H. Ritter, *Macromol. Rapid Commun.*, 2005, **26**, 160.
63. R. Hoogenboom, M. A. M. Leenen, F. Wiesbrock and U. S. Schubert, *Macromol. Rapid Commun.*, 2005, **26**, 1773.
64. F. Wiesbrock, R. Hoogenboom, M. A. M. Leenen, M. A. R. Meier and U. S. Schubert, *Macromolecules*, 2005, **38**, 5025.
65. R. Hoogenboom, F. Wiesbrock and U. S. Schubert, Abstracts of Papers, *230th ACS National Meeting*, Washington, DC, United States, Aug. 28– Sept. 1, 2005, PMSE-516.
66. H. Zhang, R. Hoogenboom, M. A. R. Meier and U. S. Schubert, *Meas. Sci. Technol.*, 2005, **16**, 203.
67. http://www.google.co.in/url?q = http://en.wikipedia.org/wiki/Copolymer.
68. X. M. Fang, R. Hutcheon and D. A. Scola, *J. Polym. Sci., Part A: Polym. Chem*, 2000, **38**, 1379.
69. L. Liu, Y. Li, Y. Fang and L. X. Chen, *Carbohydr. Polym.*, 2005, **60**, 351.
70. R. Nagahata, D. Sano, H. Suzuki and K. Takeuchi, *Macromol. Rapid Commun.*, 2007, **28**, 437.
71. Z. J. Yu and L. J. Liu, *J. Biomater. Sci., Polym. Ed.*, 2005, **16**, 957.
72. C. Zhang, L. Liao and S. Gong, *Macromol. Rapid Commun.*, 2007, **28**, 422.
73. B. Luo, C. Zhong, Z. He and R. Chang-ren, *Bioinform. Biomed. Eng.*, 2009, **1**, 978.
74. Z. Yu and L. Liu, *J. Appl. Polym. Sci.*, 2007, **104**, 3973.
75. M. I. Malik, B. Trathnigg and C. O. Kappe, *Macromol. Chem. Phys.*, 2007, **208**, 2510.
76. R. Hoogenboom and U. S. Schubert, *Macromol. Rapid Commun.*, 2008, **28**, 368.
77. N. Adams and U. S. Schubert, *Adv. Drug Delivery Rev.*, 2007, **59**, 1504.
78. M. I. Malik, B. Trathnigg and C. O. Kappe, *J. Chromatogr. A*, 2009, **1216**, 1167.
79. C. Guerrero-Sanchez, R. Hoogenboom and U. S. Schubert, *Chem. Commun.*, 2006, **25**, 3797.
80. M. Deetlefs and K. R. Seddon, *Chim. Oggi*, 2006, **24**, 16.

Microwave-assisted Peptide Synthesis

5.1 Introduction

Peptides are short chains of amino acid monomers linked by peptide bonds. Peptide bonds are the amide, covalent linkages formed when the carboxyl group of one amino acid reacts with the amino group of another. Peptides usually contain approximately 50 amino acids or fewer, the shortest being dipeptides, which consist of just two amino acids joined by a single peptide bond. A polypeptide is a long, continuous, unbranched peptide chain.[1]

The development of peptide synthesis fueled several application areas in which synthetic peptides are now used. There is a plethora of fields in which synthetic peptides now find applications. These include *immune monitoring, i.e.* the tracking of immune responses after a particular drug treatment or immunization. It is a useful way of monitoring the effectiveness of treatments.

Epitope mapping helps to identify and characterize novel epitopes (a localized region on the surface of an antigen that is capable of eliciting an immune response and of combining with a specific antibody to counter that response)[2] from a protein. This mapping in turn leads to the production of immunotherapies or vaccines for a wide range of disease areas such as cancers and other infectious diseases.

Cell signaling is yet another significant immunological application of synthetic peptides. This comprises decoding the phosphorylation/ dephosphorylation process regulated by kinases and phosphatases, respectively. Synthetic peptides are used to study enzyme-substrate interactions within important enzyme classes such as kinases and proteases, which play a crucial role in cell signaling.

RSC Green Chemistry No. 35
Microwave-Assisted Polymerization
By Anuradha Mishra, Tanvi Vats and James H. Clark
© Anuradha Mishra, Tanvi Vats and James H. Clark 2016
Published by the Royal Society of Chemistry, www.rsc.org

Synthetic peptides are used as standards and reagents in mass spectrometry (MS)-based applications. Synthetic peptides play a central role in MS-based discovery, characterization and quantitation of proteins, especially those that serve as early biomarkers for diseases.

The ever-increasing interest in synthetic peptides has generated significant research activities in the field of peptide synthesis. This chapter deals with the basics of peptide synthesis and provides a comprehensive overview of the utilization of microwave heating for peptide synthesis.

5.2 Peptide Synthesis: the Process

Peptide synthesis is usually accomplished by coupling the carboxyl group of the incoming amino acid to the N-terminus of the growing peptide chain. It is worthwhile to note that C-to-N synthesis is the opposite of protein biosynthesis, where the N-terminus of the incoming amino acid is linked to the C-terminus of the protein chain (N-to-C). During *in vitro* protein synthesis, the addition of amino acids to the growing peptide chain is performed in a precise, step-wise and cyclic manner. Despite considerable differences in the synthetic methodology, all the artificial peptides are generated by the same step-wise method to add amino acids one at a time to the growing peptide chain.

(a) Peptide Deprotection

As amino acids contain multiple reactive groups, a peptide synthesis must be strategically performed in order to avoid side reactions that can potentially reduce the length and cause branching of the peptide chain. To minimize side reactions, certain chemical groups have been developed that bind to the amino acid reactive groups to protect the functional group from unwanted reactions. These are called protecting groups (Scheme 5.1).

The amino acids used in peptide synthesis are purified first and subsequently react with these protecting groups prior to synthesis. The specific protecting groups are then removed from the newly added amino acid (a step called deprotection) immediately after coupling to allow the next incoming amino acid to bind to the growing peptide chain in the desired orientation. After the completion of peptide synthesis, all remaining protecting groups

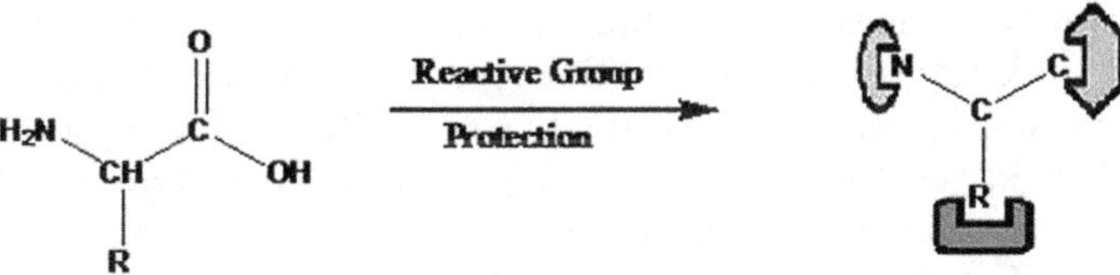

Scheme 5.1 Reactive group protection.

Table 5.1 Amino acid abbreviations.

Amino acid	Three-letter abbreviation
Alanine	Ala
Arginine	Arg
Asparagine	Asn
Aspartate	Asp
Cysteine	Cys
Glutamate	Glu
Glutamine	Gln
Glycine	Gly
Histidine	His
Isoleucine	Ile
Leucine	Leu
Lysine	Lys
Methionine	Met
Phenylalanine	Phe
Proline	Pro
Serine	Ser
Threonine	Thr
Tryptophan	Trp
Tyrosine	Tyr
Valine	Val

are removed from the initial peptide. Table 5.1 contains the list of amino acid abbreviations. Depending on the method of peptide synthesis, three main types of protecting groups are used, which are described as follows.

(i) N-terminal protecting groups: These types of protecting groups are also termed "temporary" protecting groups on account of their easy removability.

Two commonly used N-terminal protecting groups are *tert*-butoxycarbonyl (Boc) and 9-fluorenylmethoxycarbonyl (Fmoc). Each group has distinct characteristics that determine their use. On the one hand, Boc requires a moderately strong acid such as trifluoracetic acid (TFA) to be removed from the newly added amino acid, while on the other, Fmoc is a base-labile protecting group which can be removed with a mild base such as piperidine (Figure 5.1).

Boc chemistry was first described by Merrifield[3–8] and requires acidic conditions for deprotection. The Boc strategy commonly uses hydrofluoric acid (HF) for the release of assembled peptide from the support, thus requiring specialized non-corrosive HF apparatus. Fmoc strategy, which was introduced much later, requires mild, basic conditions for deprotection. It is because of mild reaction conditions, higher quality and greater yield that Fmoc chemistry is more commonly used in commercial settings. Boc strategy is preferred for complex peptide synthesis or when non-natural peptides or analogs that are base-sensitive are required.

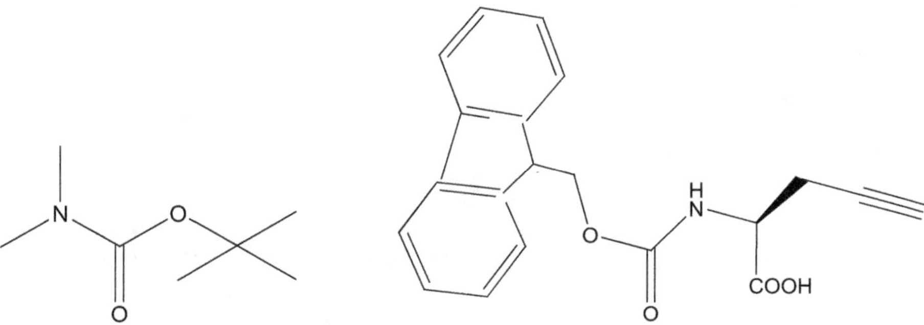

Figure 5.1   *Tert*-butoxycarbonyl (Boc) and 9-fluorenylmethoxycarbonyl (Fmoc).

(ii) C-terminal protecting group: The use of a protecting group on the C-terminal of the initial amino acid depends on the type of peptide synthesis used. The liquid-phase peptide synthesis requires protection of the C-terminus of the first amino acid (C-terminal amino acid), while its use is ruled out in solid-phase peptide synthesis because a solid support (resin) acts as the protecting group for the only C-terminal amino acid that requires protection. Amino acid side chains have multiple functional groups and are therefore prone to side-chain reactions during peptide synthesis. On account of this, protecting groups usually based on the benzyl (Bzl) or *tert*-butyl (*t*Bu) group are used. The specific protecting group to be used is governed mainly by the peptide sequence and the type of N-terminal protection used.

(iii) Side-chain protecting groups: These groups are known as permanent protecting groups, because they can withstand the multiple cycles of chemical treatment during the synthesis phase and are only removed during treatment with strong acids after synthesis is complete. The need for protection of certain side-chain protecting groups is a requisite, especially the ε-amino function of lysine, where branched peptides would result from acylation of an unprotected residue.

For a successful peptide synthesis, it is important to chalk out a protecting scheme that matches protecting groups so that deprotection of one protecting group does not affect the binding of the other groups. This is represented in Scheme 5.2. For trifunctional amino acid residues, *e.g.* Cys, Asp, Glu, Lys and Arg, side-chain protection is essential for successful peptide synthesis. The commonly used protecting groups are: *tert*-butyl (*t*-Bu) for Glu, Asp, Ser, Thr and Tyr; 2,2,4,6,7-pentamethyl-dihydrobenzofuran-5-sulfonyl (Pbf) for Arg; and trityl (Trt) for Cys, Asn, Gln and His (Table 5.2).

The act of removing protecting groups, especially under acidic conditions, results in the production of cationic species that can alkylate the functional groups on the peptide chain. Therefore, scavengers such as water, anisol or thiol derivatives can be added in excess during the deprotection step to react with any of these free reactive species.

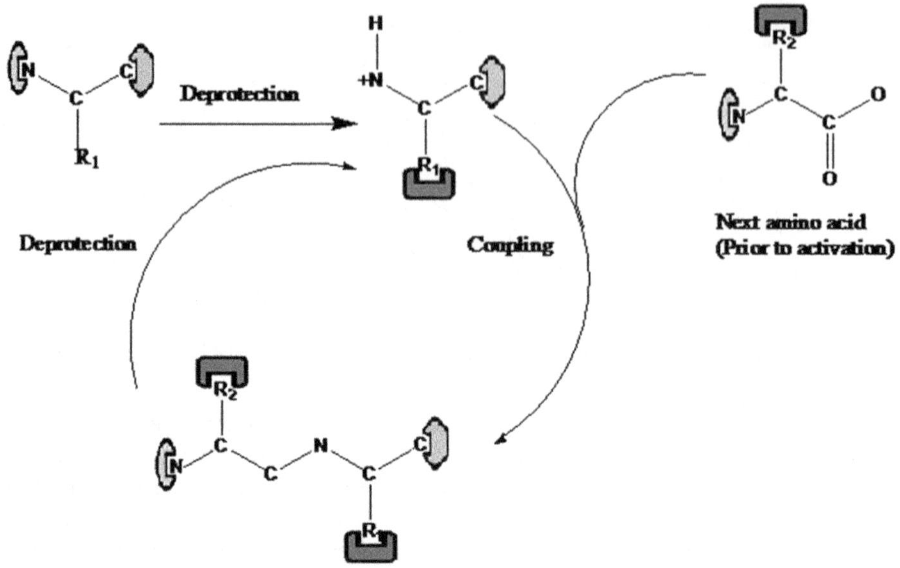

Scheme 5.2   Various steps involved in peptide synthesis.

Table 5.2   Common protecting scheme-specific solvents.

Protecting scheme	Deprotection	Coupling	Cleavage	Wash
Boc/Bzl	TFA	Coupling	HF, HBr, TFMSA	DMF
Fmoc/*t*Bu	Piperidine	Agent in DMF	TFA	

(b) Amino Acid Coupling

The C-terminal of carboxylic acid on the incoming amino acid is coupled using carbodiimides such as dicyclohexylcarbodiimide (DCC) or diisopropylcarbodiimide (DIC).[9–12] These coupling reagents react with the carboxyl group to form a highly reactive *O*-acylisourea intermediate that is quickly displaced by nucleophilic attack from the deprotected primary amino group on the N-terminus of the growing peptide chain to form the nascent peptide bond. The major drawback of carbodiimide-based reagents is the O→N rearrangement of the *O*-acylisourea intermediate and "overactivation" by formation of the symmetrical anhydride,[13] therefore, carbodiimides are used in combination with auxiliary nucleophiles such as HOBt(1-hydroxybenzotriazole) or HOAt.[14]

These auxiliary nucleophiles maintain the optical integrity of the stereogenic center at the C-terminal of the activated amino acid residue throughout the coupling step (Figure 5.2). Numerous coupling reagents have been developed to reduce coupling time and minimize epimerization, since the carbodiimide-based coupling reagents were introduced – the most

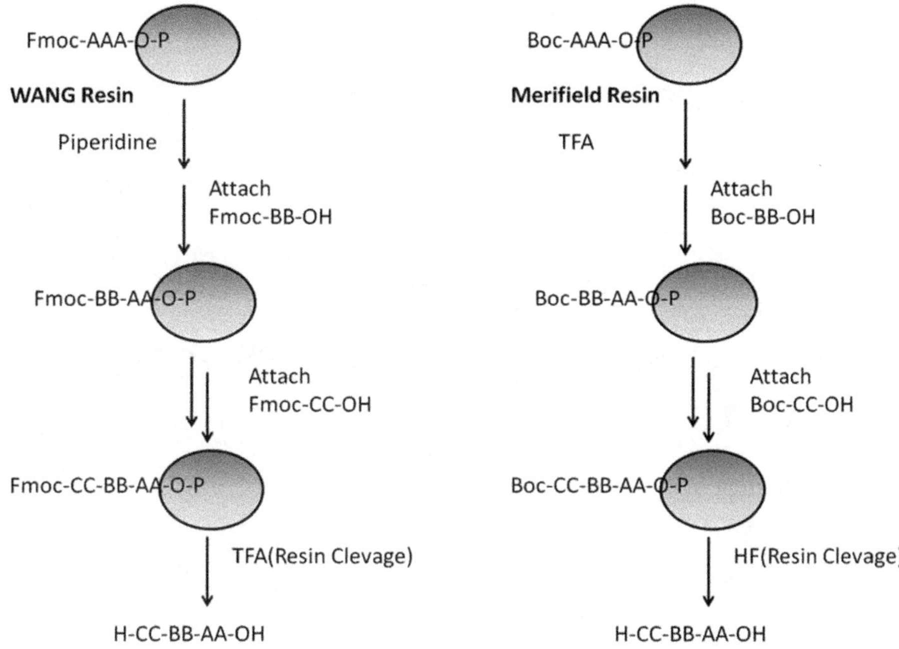

Figure 5.2 Comparative account of Fmoc and Boc strategies of peptide synthesis.

important are HBTU (N-[(1H-benzotriazol-1-yl)(dimethylamino)methylene]-N-methylmethanaminium hexafluorophosphate N-oxide),[15,16] HATU (N-[(dimethylamino)-1H-1,2,3-triazole[4,5-b]pyridine-1-ylmethylene]-N-methylmethanaminium hexafluorophosphate N-oxide),[17] PyBOP (1-benzotriazolyloxy-tris-pyrrolidinophosphonium hexafluorophosphate)[16] and the novel COMU (1-[(1-(cyano-2-ethoxy-2-oxoethylideneaminooxy)-dimethylaminomorpholino-methylene)]methanaminium hexafluorophosphate)[18] reagents (Figure 5.3).

(c) Peptide Cleavage

Post completion of successive cycles of amino acid deprotection and coupling, all remaining protecting groups are removed from the nascent peptide by acidolysis. Strong acids such as hydrogen fluoride (HF), hydrogen bromide (HBr) or trifluoromethane sulfonic acid (TFMSA) are used to cleave Boc and Bzl groups, while a relatively milder acid such as TFA is used to cleave Fmoc and tBut groups. This cleavage should ideally result in the removal of the N-terminal protecting group of the last amino acid added, the C-terminal protecting group from the first amino acid and any side-chain protecting groups. Scavenger molecules are also included during this step to react with free protecting groups. The major challenge in this step lies in the optimization of the process to avoid any acid-catalyzed side reactions. This is pictorially demonstrated in Scheme 5.3.

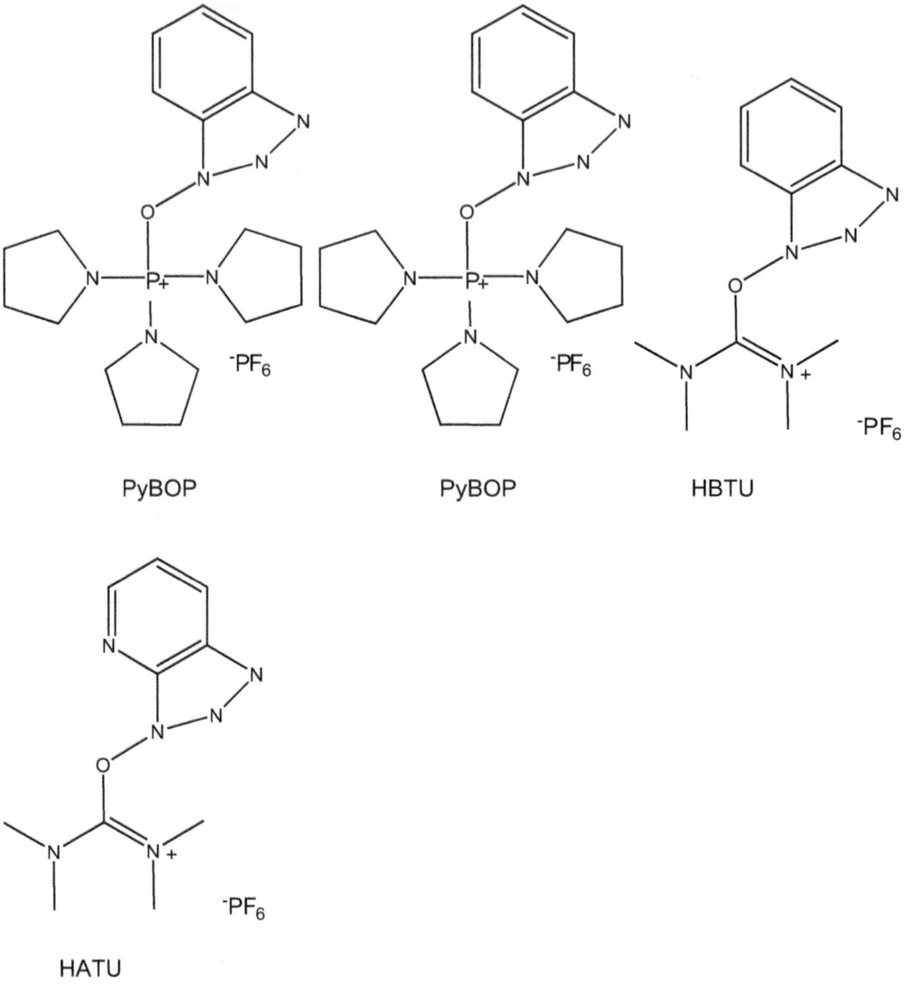

PyBOP   PyBOP   HBTU

HATU

Figure 5.3   Some common coupling agents used in peptide synthesis.

5.2.1   Peptide Synthesis Strategies

The two synthetic strategies for peptide synthesis are:

(a) Liquid-phase peptide synthesis.
(b) Solid-phase peptide synthesis.

Liquid phase synthesis is the classical method of synthesis and is still commonly used for large-scale synthesis. The product has to be manually removed from the reaction solution after each step, making it slow and labor-intensive. Apart from this, the requirement of another chemical group to protect the C-terminus of the first amino acid is also a drawback. The fact

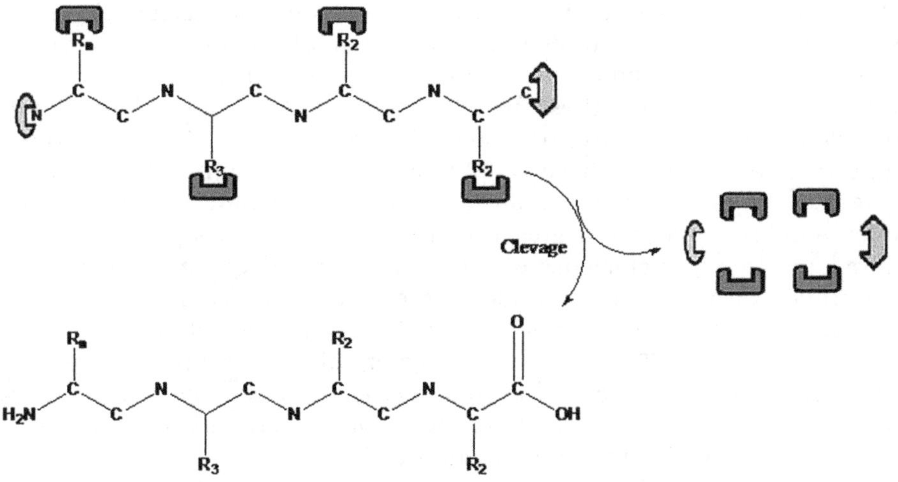

Scheme 5.3   Peptide cleavage after synthesis.

that in the liquid-phase synthesis the product is purified after each step makes the detection side reactions easy. In addition, convergent synthesis can be performed in the liquid phase in which separate peptides are synthesized and then coupled together to create larger peptides.

Solid-phase peptide synthesis (SPPS), which was pioneered by Robert Bruce Merrifield,[3] has now become an accepted method for creating peptides and proteins in the lab in a synthetic manner. SPPS allows the synthesis of natural peptides, which are difficult to express in bacteria, through the incorporation of unnatural amino acids. In this method, instead of C-terminal protection with a chemical group, the C-terminus of the first amino acid is coupled to an activated solid support, such as polystyrene or polyacrylamide. This type of approach has a two-fold function: the resin acts as the C-terminal protecting group and provides a rapid method to separate the growing peptide product from the different reaction mixtures during synthesis. As with many different biological manufacturing processes, peptide synthesizers have been developed for automation and high-throughput peptide production.

Over time, peptide synthesis has witnessed considerable improvement in terms of amino acid protecting groups, coupling reagents and resins. With the introduction of commercially available automated peptide synthesizers, the process has developed a high degree of predictability and reproducibility. With the current rate of development it is even predicted that direct synthesis of small proteins by SPPS will soon be possible. Drawbacks of this technique, however, include low purities and at times even failure to achieve the desired peptide sequence. Amino acids, which are prone to forming β-sheets, lead to aggregation during peptide synthesis because of their hydrogen bonding and hydrophobic properties.[19–24] The aggregation of peptides very often leads to premature terminations or deletions of the elongating peptide sequence. This leads to a mixture of aggregated peptide sequences, which is very difficult to separate and leads to poor solvation.

In order to overcome this aggregation, researchers have come up with several solutions. These include use of resins with a low-loading, pseudoprolines,[25] composition[26] and chaotropic salts.[27] These, however, have limited usage and have variable efficiencies. Heating, however, has emerged[27–32] as an additional aid in SPPS and is helpful in reducing both inter- and intramolecular aggregation, thereby reducing coupling times and improving coupling efficiency of bulky and β-branched amino acids. The application of heat during peptide synthesis, however, comes with its own challenges, as peptides are polyfunctional moieties and can degrade at elevated temperatures. The fact that in a biological context, proteins are functional at temperatures well above 40 °C has generated an enormous interest in the application of heat in SPPS strategies, which were earlier considered to be developed for use at room temperature.

Slowly and gradually, research involving the use of conductive heating in various steps of peptide synthesis started gaining grounds.[27–32] In 1997, Varanda and Miranda reported the synthesis of the acyl carrier protein fragment 65–74 (ACP(65–74)) and the un-sulfated cholecystokinin-8 at 60 °C (temperature used in the coupling steps).[32] Kaplan and co-workers synthesized long peptides (84–107 residues) using elevated temperatures both in the N^α-deprotection (40 °C) and the coupling (55 °C) steps.[29]

After it was established that conductive heating increases coupling yields, the use of microwave heating in peptide synthesis attracted researchers from all over the world. In the following section, we will be dealing with the effects and applications of microwave irradiations in SPPS.

5.3   Microwaves in SPPS

Wang and co-workers were first to report SPPS of ACP(65–74) as well as two other peptides using a slightly modified domestic microwave oven.[33,34] Erdelyi and Gogoll demonstrated that special microwave heating not only improved speed and purity in SPPS, it also prevented the degradation of solid support.[35] Post this, application of microwave heating to direct and linear SPPS methodology has resulted in the synthesis of a considerable number of peptides and proteins up to at least 109 amino acid residues.[36]

Microwave heating (dielectric heating at 2.45 GHz) occurs by disposing the energy directly to the solvent (and some reagents), due to interactions of the material with the alternating electric field. We can thus conclude that the optimization of solvent plays a major role in any kind of microwave synthesis. *N,N*-Dimethylformamide (DMF) and *N*-methyl-2-pyrrolidinone (NMP) are the most common solvents used for both coupling and N^α-deprotection in SPPS.[37] Apart from the solvent, sample volume, vessel material and the mode of stirring, *i.e.* vortexing or N_2 bubbling have a profound impact on the reaction profile. The major drawback of microwave-assisted SPPS is inhomogeneous temperature distributions in the reactor vessel. The solvents, which strongly absorb microwave energy, give short microwave penetration and the addition of resin may increase solution viscosity.

Several groups propose that microwave radiations to a significant extent interact directly with amide dipoles of the peptides, and that this effect results in direct heating of peptides, as opposed to indirect heating by contact with the solvent molecules. This explains why microwave heating in SPPS is not only faster, but also provides higher purities compared to conventional room temperature SPPS.[38-40] Bacsa *et al.* compared microwave heating with conventional heating during SPPS.[37] They tested three peptides, varying in length from 9 to 24 AA, prepared by microwave and oil-bath heating at 86 °C on two different resins, which resulted in similar crude purities. Moreover, the degree of epimerization and the impurity profiles were identical for the two peptides. Finally, the 60 °C increase in temperature (ambient to 86 °C) leads to an estimated 50-fold increase in reaction rate from both processes, in agreement with the Arrhenius equation, thus this kinetic effect is probably responsible for providing peptides at high speed and purity.[37]

In the upcoming segments, we will discuss the application and extent of microwave irradiations in the various stages of peptide synthesis.

5.3.1 Peptide Couplings

The use of microwave-irradiations in Fmoc and Boc strategies has shown an increase in reaction rates and crude peptide purity, though Fmoc strategy is the more common of the two. Amongst the coupling reagents, almost all showed an increased efficiency under the MW irradiations, barring the auxiliary nucleophile Oxyma (ethyl 2-cyano-2-(hydroxyimino)acetate), where the thermal stability has been reported to be relatively low when subjected to heating.[41] The majority of controlled microwave-assisted peptide couplings were performed at temperatures in the range 50–80 °C, but for amino acids with an increased risk of epimerization, which often were coupled at temperatures under 50 °C.[38]

Systematic studies where coupling reagents, temperature and time were varied have been published.[36] These studies report that the coupling agents that worked well with conventional SPPS were equally useful when applied to microwave-assisted SPPS. It has been reported that the "uranium" type of coupling reagent, COMU, was considerably better than the classical HBTU and HATU in the synthesis of an Aib-containing pentapeptide,[41] and PyBOP, DIC, TSTU, HCTU, HBTU as well as HATU in the synthesis of the C-terminus of the MuLV CTL epitope.[42,43] Typically, the amino acids and coupling reagents were used in 3–5 molar excess, however, for difficult sequences up to 10 molar excess has been reported.

There are several reported cases where intermediate cooling was applied to suppress different side reactions and epimerization. These reactions were mainly performed on a manual microwave instrument, the Discover SPS (CEM). The reaction mixture was cooled up to 10 °C in an ice–ethanol solution, followed by irradiation with 100 W for 5 s, which gave a temperature of approximately 30 °C, and the cycle was repeated 5 times.[44–47] It is, however, worthwhile to note that no systematic studies have been reported wherein detailed effects of intermediate cooling has been explained.

A few important case studies where microwave irradiations have been successfully applied in the synthesis of long peptide sequences are described in the following segment. Synthesis of human exon 1 huntingtin, which consist of a 42 residue poly-Gln stretch and two poly-Pro stretches of 10 and 11 residues, respectively, connected by spacer regions, giving a total of 109 amino acid residues,[48] holds the current record of longest peptide sequence assembled by microwave-assisted SPPS. β-amyloid(1–42), which is considered to be an example of a long and difficult sequence was successfully synthesized using microwave heating.[49,50] The complex peptide hormone insulin (containing two peptide chains and three disulfide bridges) has been synthesized by conventional SPPS using complex strategies.[51,52] The application of MW heating allowed the synthesis of desB30 insulin analogs in reasonable yields through the assembly of a 60-mer linear precursor, which on folding was processed to two-chain insulin.[53] In another study, a 58-*mer* peptide (an affibody) composed of three a-helices was successfully synthesized using CEM standard conditions.[54] The microwave irradiations under these particular conditions led to an increase in deletion sequences as well as an increase in aspartimide formation compared to room temperature. Another affibody, a 66-mer peptide, was synthesized in 41% crude purity using microwave-assisted SPPS.[55]

5.3.2 Phosphopeptides

As synthetic phosphopeptides are important in determining the process of protein phosphorylation/dephosphorylation which in turn governs a myriad of cellular processes, the development of efficient solid-phase synthesis methods is an important research area. The synthesis is generally performed by incorporation of phosphorylated amino acid derivatives into the growing peptide chain.[56] Owing to their high chemical stability upon storage over long periods, Fmoc-Xxx(PO$_3$Bn,H)–OH (Xxx = Ser, Thr or Tyr) have proven particularly useful in Fmoc. In this procedure the phospho-protecting group is removed during the final TFA treatment. The room temperature introduction of phosphorylated derivatives often requires double coupling,[57] however, microwave heating is said to increase the yield of the incorporation of the phosphorylated amino acid as well as shorten the coupling time.[58–60] Attard and co-workers reported an increased piperidine induced phosphoryl β-elimination of Ser(PO$_3$Bn,H)-containing peptides at elevated temperature.[61] They also reported that the use of 50% cyclohexylamine in DCM, 5% DBU in DMF or 5% piperazine in DMF instead of 20% piperidine in DMF for the N$^\alpha$-deprotection of Fmoc-Ser(PO$_3$Bn,H)-peptidyls eliminated several β-elimination problems. This was attributed to the fact that piperidine-induced β-elimination occurs only for phosphoseryl residues, not for Tyr, and most likely only during the N$^\alpha$-deprotection of the Ser(PO$_3$Bn,H) residue.[61] Brandt *et al.* used monobenzylated phosphorylated amino acid building blocks to demonstrate that microwave irradiation both during coupling and N^{α}-deprotection provided phosphopeptides in moderate purity.[59]

5.3.3   *O*- and *N*-glycopeptides

Glycopeptides are those peptides which contain carbohydrate moieties (glycans) covalently attached to the side chains of the amino acid residues that constitute the peptide. The glycan moiety of glycopeptides plays a crucial role in several vital biological functions, such as fertilization, the immune system, brain development, the endocrine system and inflammation. The synthesis of glycopeptides provides biological probes for elucidation of glycan function in nature and leads to the development of the products that have useful therapeutic and biotechnological applications.[62] Some important glycosylated peptides synthesized by microwave-assisted reactions are given in Figure 5.4.

Several methods have been reported in the literature for the synthesis of glycopeptides. Of these methods, the most common strategies are as follows.

(a) Solid Phase Peptide Synthesis (SPPS)

Within SPPS, there exist two strategies for the synthesis of glycopeptides, linear and convergent assembly. Linear assembly relies on the synthesis of building blocks and then the use of SPPS to attach the building block together.

(b) Native Chemical Ligation (NCL)

Native chemical ligation, or NCL, is a convergent synthetic strategy based on the linear coupling of glycopeptide fragments. This technique makes use of the chemoselective reaction between a N-terminal cysteine residue on one peptide fragment with a thio-ester on the C-terminus of the other peptide fragment.

Glycopeptides are commonly synthesized by incorporating pre-synthesized glycosylated amino acid building blocks[63–67] into stepwise SPPS. The major advantages of microwave-assisted solid-phase glycopeptide synthesis (SPGS) are short and efficient coupling (which reduces the cost as well as reaction time) even for complex and long glycosylated amino acid building blocks. However, the major limitation of using microwave irradiation is destabilization of the glycosylated building block during the coupling step and N^{α}-deprotection. The piperidine treatment during N^{α}-deprotection may lead to β-elimination of the glycan moiety on application of microwave irradiations.

Microwave-assisted SPGS is widely reported in mucin-type glycopeptides.[68] Nishimura and co-workers have used microwave-assisted SPGS to synthesize a variety of MUC1-related glycopeptides (incorporation of glycosylated amino

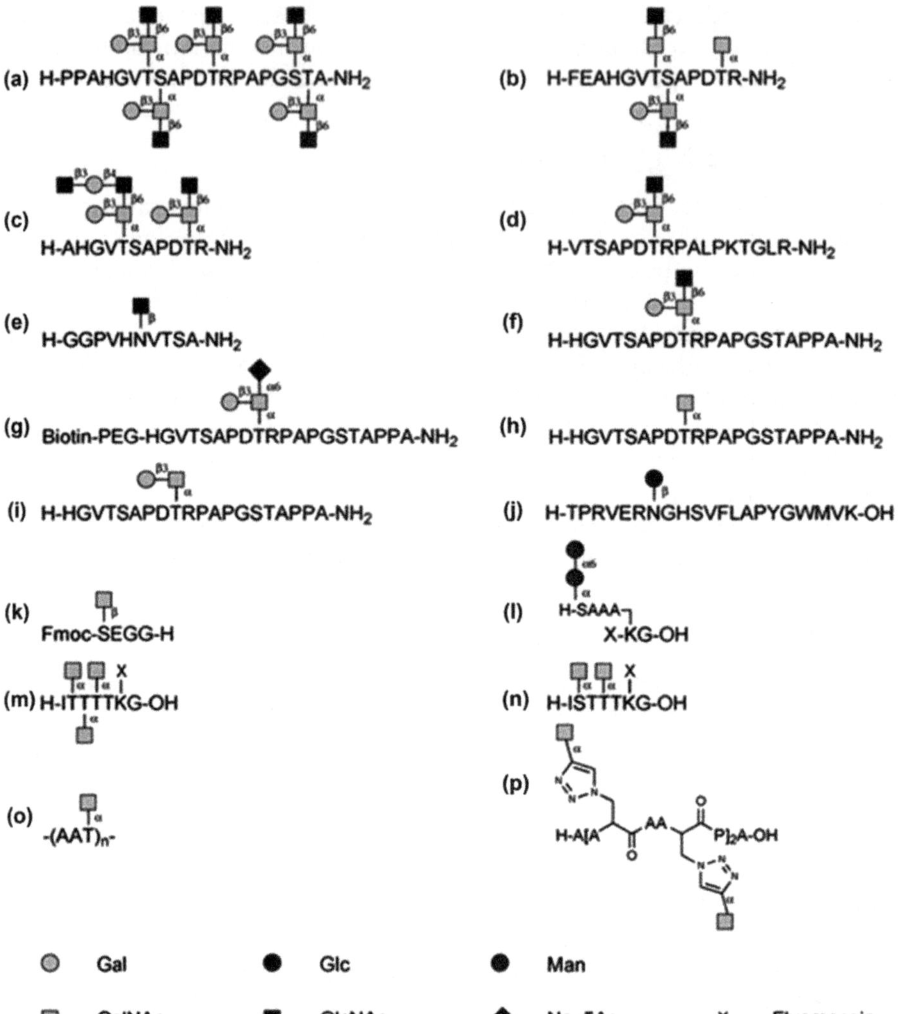

Figure 5.4 Glycosylated peptides synthesized by microwave-assisted SPGS.[36]

acids).[69–74] Commonly incorporated building blocks in MUC1 type glycopeptides are a core-2 trisaccharide on Ser or Thr but other saccharides have also been incorporated.[71–74] The Nishimura group reported microwave-assisted incorporation of O-linked pentasaccharide at 50 °C for 20 min. It is the most complex glycosylated building block incorporated by microwave-assisted SPGS to date.[72] The same group reported the microwave-assisted synthesis of the most complex glycopeptide that contained five core-2 trisaccharide on Ser and Thr residues and the overall synthesis time required for the synthesis of this 20-mer was reduced from 4 days to 7 hours, with comparable overall yields.[69,70] This solid-phase strategy was used to synthesize enzymatically elongated (post-SPGS) glycopeptides by glycosyltransferases.[71–74] Nishimura and co-workers incorporated a N-linked Asn(Ac3-bGlcNAc) residue into a 10-*mer* glycopeptide and further elongated it enzymatically.[73]

Several groups have reported that, during the synthesis of CSF114, (which is a *N*-glycosylated peptide), the incorporation of the glycosylated building block benefitted from the increased temperature and shorter coupling time.[74–76] The Papini group performed the incorporation N-linked Asn(Ac$_4$-βGlc) moiety at 70–75 °C for 5 min and the N$^\alpha$-deprotection at 70–75 °C for 3–3.5 min.[75] The application of microwave irradiation not only led to a reduction in cycle time (from 2 h to 30 min), it also increased the crude peptide purity, as well as overall yield (from <20% to >70% and 10% to 46%, respectively). This strategy was used in the synthesis of variety of glycopeptides with a N-linked Asn(βGlc) residue.[77]

Brimble and co-workers devised a methodology for the efficient, automated and microwave-assisted Fmoc solid-phase synthesis of a 5(6)-carboxyfluorescein-labelled Lys(Dde)-Gly-Wang resin. They incorporated carboxyfluorescein-labelled *O*-dimannosylated peptides onto this resin using Fmoc-[α-d-Man(OBz)$_4$-(1 → 6)-α-d-Man(OBz)$_3$α1-]Ser-OH and PEG-[α-d-Man(OBz)$_4$-(1 → 6)-α-d-Man(OBz)$_3$α1-]-OH building blocks. These *O*-dimannosylated glycopeptides proved very important for biological evaluation, fluorescently labeled glycopeptides containing GalNAcα1-*O*-Ser/Thr residues provide valuable immunological probes for the development of cancer vaccines. This synthetic methodology is flexible and suitably robust for the incorporation of three contiguous GalNAcα1-*O*-Ser residues into the peptide chain. This was attained by applying high temperatures (80 °C) during the coupling (20 min) of the glycosylated building blocks as well as during N$^\alpha$-deprotection (0.5 + 3 min).[58]

Antifreeze proteins or ice structuring proteins refer to a class of polypeptides produced by certain vertebrates, plants, fungi and bacteria that permit their survival in subzero environments. Structurally, they are composed of the tripeptide units (Ala–Ala–Thr)*n* with minor sequence variation and bear an O-linked β-Gal-(1–3)-αGalNAc at every Thr residue. Several groups have synthesized antifreeze glycopeptides and analogs by microwave-assisted SPGS.[78–80] Sewald and co-workers applied microwave irradiations to incorporate Thr residue with an O-linked α-GalNAc (using SPGS) with a maximum temperature of 40 °C. The procedure reduced the cycle time from 3 h to 45 min.[78] The procedure at higher temperatures resulted in a decomposition and de-glycosylation of the peptides.[78] In contradiction with the above results, the data for the microwave-assisted SPGS of an antifreeze peptide analog containing four neoglycosylation sites where the coupling step was performed at high temperature (80 °C) over 15 min, decomposition or de-glycosylation of the target peptide was not reported.[79]

Unverzagt and co-workers worked on fragment condensation of RNase 39-*mer* glycopeptide thioesters. Three fragments were coupled onto a resin-bound glycopeptide using microwave irradiation for 2×30 min at 55 °C. The glycopeptides, having an unprotected GlcNAc moiety, were not reported to decompose or de-glycosylate, *i.e.* after 3 hours of microwave heating at 55 °C.[81]

The area of microwave heating during SPGS is not a highly explored field, but going by the current data it has been suggested that heating can improve the yields of synthesized peptides and minimize the reaction times. At the same time, decomposition and de-glycosylation at higher temperatures cannot be completely ruled out and only a small section of glycosylated amino acids have been incorporated using this technique.

5.3.4 N^{α}-methylated Peptides

N-methylation is one of the simplest chemical modifications that can be performed in peptides and proteins. The process comes across as an ingenious, natural technique to modulate biological function, often as a mode of survival through the production of antibiotics. Methylation is a small structural change, that can not only mobilize large protein complexes (as in the histone methylation), but also acts as enzyme inhibitor by selective recognition of protein–protein interaction surfaces. Of late, due to the well-researched synthetic approaches, the potential of *N*-methylation has been established in modulating biological activity and selectivity, pharmacokinetic properties of peptides, and also in delivering novel drugs.[82–88]

The *N*-methylated peptides are usually synthesized by a three-step procedure.[89] In the first step, the amine is N-terminally methylated using dimethylsulfate as the methylating agent,[90,91] or by incorporating Fmoc *N*-methyl amino acid building blocks by the usual SPPS procedure.[92–96] The subsequent coupling onto *N*-methyl amino acids is often done using coupling reagents such as triphosgene, HATU or PyBOP.[92–96] Up to now, the use of the microwave heating for the assembly of *N*-methylated peptide is relatively unexplored. Albericio and co-workers pioneered in the application of microwaves for the synthesis of short peptides containing several *N*-methyl amino acids (35 °C, 20 min) using DIC/HOAt as coupling reagents, in DCM.[97] Coupling onto highly sterically demanding residues is very troublesome and microwave heating reportedly increases the yield of these couplings tremendously.

5.3.5 β-Peptides

β-peptides are made up of β-amino acids, which have their amino group bonded to the β-carbon rather than the usual α carbon. They belong to a class of foldamers which can adopt a variety of secondary structures. β-peptides are proteolytically and metabolically stable and hence very useful for biomedical applications.[98–106] One of the prime limitations in the application of conventional SPPS conditions towards β-peptide synthesis is to attain high yields in amide bond formation and *N*-deprotection due to aggregation and folding of the peptidyl intermediates. In contrast to the extensive application of microwave irradiation in the synthesis of α-peptides, as well as for increasing the reaction rate, only a few protocols using microwave heating have been reported for the solid-phase synthesis of β-peptides.

Murray and Gellman were the first to report a microwave-assisted application for the assembly of 14-helical β-peptides, by using β^3-amino acids and trans-2-aminocyclohexane carboxylic acid (ACHC, Figure 5.5).[106–108] They synthesized a hexa and a deca-β-peptide at room temperature using double couplings and double *N*-deprotections of the ACHC residues (Figure 5.5). The hexa-β-peptide was reported to have a moderate crude purity of 55% in contrast to the penta-β-peptide precursor that had a high crude purity of 95%.[106] Application of microwave irradiation to the synthesis of the hexa-β-peptide, during the coupling (60 °C, 2 min) and *N*-deprotection (50 °C, 4 min) steps, resulted in an improved purity of 80%. The deca-β-peptide sequence under conventional room temperature SPPS conditions only gave a crude purity of 21%, however, applying microwave heating resulted in a crude peptide purity of 57%. Incomplete coupling of the ACHC residues resulted in low yields of the two β-peptides, which was addressed by performing the microwave-assisted couplings in a 0.8 M LiCl in NMP affording an

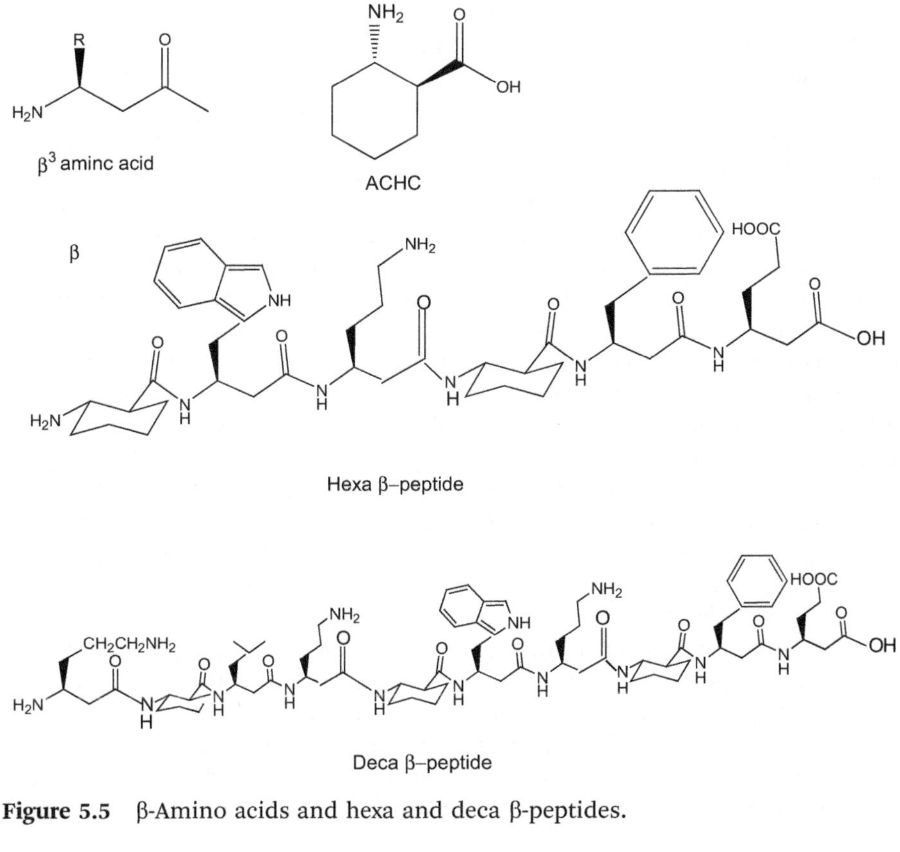

Figure 5.5 β-Amino acids and hexa and deca β-peptides.

improvement in purity to 88% and 94% for the deca-β-peptide and hexa-β-peptide, respectively.[106]

Pomerantz *et al.* and other groups used multimode microwave irradiation for the synthesis of β-peptides with a diversity of applications, such as the formation of lyotropic crystals and the synthesis of β-peptides on gold surfaces.[109–112] Petersson and Schepartz reported an optimized protocol for the synthesis of the 28-*mer* β-peptide Z28, which on using PyAOP/HOAt as coupling reagents instead of PyBOP/HOBt gave an increase in isolated peptide yield, from <1% up to 19%, (60 °C for 6 min, followed by cooling for 5 min at room temperature), as well as 20% piperidine in DMF followed by two times 2% DBU in DMF for *N*-deprotection (70 °C for 4 min followed by cooling for 5 min at room temperature).[113] The latter results elucidate the improved synthesis of especially long sequences of β-peptides on application of microwaves, which otherwise would be difficult to obtain in reasonable purities and yields using conventional methods.

5.4 Peptoids

Peptoids are poly-*N*-substituted glycines, where the side-chain is connected to the nitrogen of the peptide backbone, instead of the α-carbon, as in peptides. It is worthwhile to observe that the peptoids lack the amide hydrogen, which is responsible for many of the secondary structure elements in peptides and proteins. Peptoids, such as D-Peptides and β-peptides, are completely resistant to proteolysis, and have therapeutic applications where proteolysis is a major issue. The lack of hydrogen bonding in secondary structures in peptoids gives them an edge as they are not typically denatured by solvent, temperature or chemical denaturants such as urea. The lack of both chiral centers and hydrogen bond donors, as well as the conformational heterogeneity arising from tertiary amide isomerism, complicates the design of well-folded peptoid motifs. Therefore, the understanding of the intramolecular interactions that direct folding is still at an early stage, in comparison to other foldamer systems such as β-peptides. Peptoids have been developed as candidates for a range of different biomedical applications,[114,115] which include antimicrobial agents and synthetic lung surfactants,[116] as well as ligands for various proteins including Src Homology 3 (SH3 domain),[117] Vascular Endothelial Growth Factor (VEGF) receptor 2,[118] and antibody Immunoglobulin G biomarkers for the identification of Alzheimer's disease.[119] Peptoids, for the aforementioned properties, have potential applications in nanotechnology,[120] an area in which they may play an important role. The first demonstration of the use of peptoids was in screening a combinatorial library of diverse peptoids, which yielded novel high-affinity ligands for 7-transmembrane G-protein-couple receptors.[121] The fact that for the amino portion of the amino acid side chain of the peptoids any amine can be used has a very valuable commercial advantage. Thousands of commercially available amines can be used to generate unprecedented chemical diversity at each position at costs far lower than would be required for similar peptides or peptidomimetics. To date, at least 230 different amines have been implimented as side chains in peptoids.

Peptoids are often synthesized *via* the solid-phase submonomer method developed by Zuckermann *et al.*[122] This method consists of two steps: (a) acylation; and (b) dissociation. Acylation is performed by addition of bromoacetic acid (a haloacetic acid) activated by diisopropylcarbodiimide (DIC) to the amine of the previous residue. The displacement step is a typical S_N2 reaction in which an amine displaces the halide to form the *N*-substituted glycine residue (Scheme 5.4a).[123] The advantage of the submonomer approach is that it allows the use of any commercially available or synthetically accessible amine, with great potential for combinatorial chemistry. The major issue is the long reaction times per residue at room temperature, (almost up to 3 h, which can exceed in the synthesis of longer peptoids and also during the incorporation of amines with low reactivity).

Kodadek and co-workers reported microwave-aided solid-phase synthesis of peptoids using submonomer approach. Even by performing microwave experiments in a domestic microwave oven without adequate control of the temperature, they obtained 10-fold reduced synthesis time for a 9-residue peptoid.[124,125] The microwave method generally gave higher yields and purity than for those synthesized at room temperature.

Blackwell and co-workers used a laboratory microwave reactor with some temperature and pressure control (Milestone Ethos Microsynth microwave), to elucidate the effect of microwave irradiation on peptoid synthesis. They studied and concluded that incorporation of electronically deactivated benzyl amide side-chains into peptoids was significantly improved by microwave irradiation.[126,127] They compared the synthesis of a pentamer of 1-(pentafluorophenyl)ethylamine (fpe) units was at room temperature and after microwave application. At room temperature, a crude purity of 22% was obtained, however, on application of microwave irradiations, the crude purity was improved to 56%. They also inferred that microwave irradiation is not always required for high yields, and peptoids which consist of only unhindered primary amines can be synthesized, even at room temperature.[127,128] The Messeguer group also devised a microwave-assisted protocol using the "tea bag" approach, which enabled them to identify new trypsin inhibitors.[128]

Blackwell and co-workers presented an extensively studied folding mechanism of more than 38 peptoids,[129] of which six were assembled using a microwave-assisted solid-phase method.[126] In another study by the same group, synthesis of peptoid nonamers containing nitroaromatic monomer using microwave irradiation was reported. They studied the course of reaction by circular dichroism spectroscopy.[130] The twist in the reaction strategy was that the previously established microwave methodology resulted in a low yield and poor crude purity for this reaction. The low yields were attributed to the decreased nucleophilicities of these a-chiral nitro aromatic amines, which in turn led to the development of an optimized protocol in which the amination reaction time was increased from 90 s to 1 h

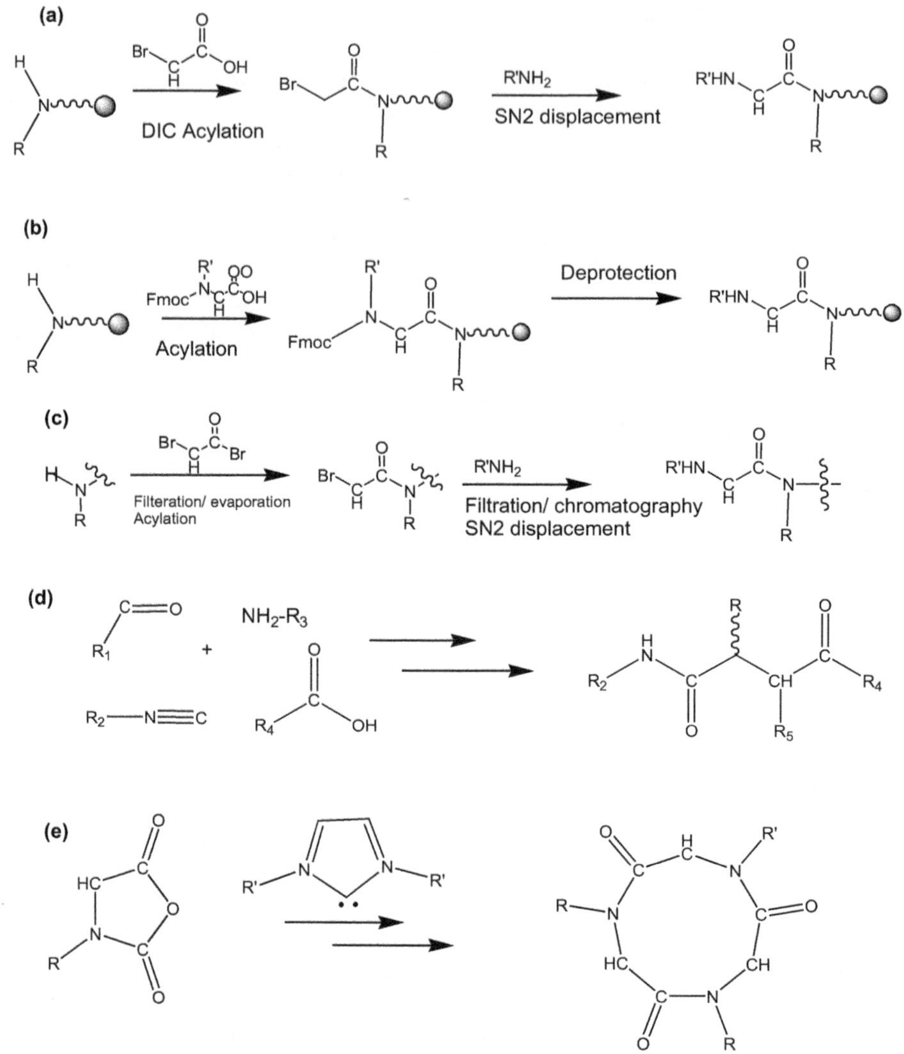

Scheme 5.4 Different methods of peptoid synthesis: (a) SPPS: Submonomer approach; (b) SPPS: Fmoc method; (c) Solution-phase synthesis: bromo-acetylbromide method; (d) Solution-phase synthesis: using four component reaction; (e) Solution-phase synthesis: amine-initiated ring-opening polymerization.[123]

and the temperature was lowered from 95 °C to 60 °C, which increased the yield from 13% to 20%.[130]

The aforementioned peptoids were synthesized using the classical Zuckermann method for peptoid synthesis.[122] An alternative solid-phase route using Fmoc peptoid building blocks was developed by Peretto *et al.* for the synthesis of peptoid oligomers.[131] The monomer unit *N*-[6-(*tert*-butyloxy-carbonyl) amino-hexyl]-*N*-Fmoc-glycine, a secondary amine, necessitates multiple couplings using PyBrOP as coupling reagent. Adding another link to the chain, Bradley and co-workers presented a microwave-assisted coupling protocol with DIC/HOBt, here the secondary amine was coupled for 20 min at 60 °C using three equiv. of monomer and the *N*-deprotections were performed using 20% piperidine in DMF (2 × 15 min) at room temperature.[132] The protocol led to the synthesis of hepta-peptoid with high purity (98%).

Several reports have been published explaining the use of microwave irradiation to synthesize labeled peptoids with a variety of fluorophores and fluorescence quenchers. These peptoids are not only ideal to study the protease activity but also make fluorescein-tagged peptoids with cell-penetrable and intracellular ability. They are also widely used for the synthesis of peptoid dendrimers.[133–135] The microwave irradiation protocols provided the peptoids in less time and in much higher crude purities.

5.5   Pseudopeptides

Apart from backbone modifications like *N*-methylated peptides, peptoids and β-peptides, pseudopeptides have also been synthesized using microwave irradiation. Pseudopeptides comprise amide bond modifications. They are applied in related studies into the importance of particular peptide bonds.[136–140] Pseudopeptides which contain reduced amide bonds are generally obtained by reductive amination of the growing peptide with Boc- or Fmoc-protected aminoaldehydes. It was in 2004 that microwave irradiation was first applied in all the steps for the synthesis of pseudopeptides. Santagada and co-workers applied microwave irradiations for the synthesis of $\psi(CH_2NH)$, both in the conversion of amino acids to aldehydes and in the reductive amination. It was a solution phase reaction.[141] Lee and co-workers subsequently reported the use of microwave heating in a solid phase synthesis of reduced amide bond surrogates. The group reportedly prepared dipeptides by reacting an aldehyde with of 1% acetic acid to give an imine, which was reduced to an amine bond using a mild reducing agent $(NaBH_3CN)$. In order to prevent the elimination of the Fmoc group, the reaction was performed at a temperature below 80 °C. The microwave irradiated reaction led to an increase in coupling yields as compared to those achieved at room temperature. The synthesis time was reduced from 5 h to 8.5 min per cycle, at room temperature and with microwave heating, respectively. The same group synthesized a 5-*mer* and a 12-*mer* pseudopeptide, having reduced amide bonds, and obtained a significant enhancement in yields compared to the unheated method (from 10 to 80%) besides a major decrease in reaction time.[142]

Lee and co-workers[143] studied the microwave-assisted synthesis of pseudopeptides containing ester bonds. These pseudopeptieds can be used to investigate the role of hydrogen bonding of amides in proteins and peptides.[144–147] They synthesized different pseudodipeptides by microwave-assisted solid-phase chemistry using (S)-2-hydroxy-4-methylpentanoic acid (a-hydroxy-Leu), DIC, DMAP (0.1 equiv.) and *N*-ethylmorpholine (NEM) (1.2 equiv.) for 12 min at 90 °C. The experiment resulted in almost identical yields as those obtained at room temperature, but the overall coupling time was reduced from 7 h to 12 min.[144] From the above results, it can be very well established that microwave protocol is an efficient and faster method of pseudopeptieds synthesis.

5.6 Microwave Irradiation in Peptide Synthesis: Potential Side Reactions

Although microwave heating is a proven and established efficient method used during the coupling step of protein synthesis, several groups have reported the use of microwave heating during the N^α-deprotection step.[38,53,148–159] The typical approach for N^α-deprotection in Fmoc chemistry involves a two-step procedure which comprises a 2–3 min treatment to remove the initially formed high concentrations of dibenzofulvene, followed by filtration and a second 10–20 min reaction to complete the N^α-deprotection. Microwave heating uses the same approach and decreases the reaction times to approximately $0.5 + 3$ min at 37–80 °C.[38,151,152,155] The major issue posed during the application of microwave heating at the N^α-deprotection step is the increased probability of side-reactions, such as aspartimide formation or epimerization. The trick to lower the chances of side reactions is to employ different reagents for Fmoc removal, for instance, 5% piperazine or 20% piperidine solutions with 0.5 M HOBt in DMF, which is strongly recommended for Fmoc removal of "difficult peptide sequences".[38,148,152,154,159,160] Another approach is intermediate cooling, which comprises 5 s cooling in between 5×5 s of 100 W microwave heating cycles.[161–164]

5.6.1 Epimerization

Microwave heating, on one hand, can accelerate the rate of amide bond formation and N^α-deprotection, but simultaneously also carries the potential to accelerate the competing side-reactions such as epimerization and aspartimide formation, thus complicating the whole synthesis mechanism. Epimerization during standard Fmoc SPPS is a common occurrence, however, during coupling and N^α-deprotection, it is not a very probable reaction. Chiral amino acids have a potentially acidic hydrogen atom at the α-carbon, which represents a potential site for epimerization through enolization reaction. The reaction involves formation of an oxazolone intermediate by the attack of the activated carboxyl group (formed by the removal of acidic, α-hydrogen atom) on the adjacent amide bond (Scheme 5.5).[165] Epimerization during peptide chain assembly is a risk when incorporating Cys or His residues.[166,167]

Epimerization can be prevented or at least reduced by reducing the chance of α-hydrogen removal. This in turn can be accomplished by avoiding

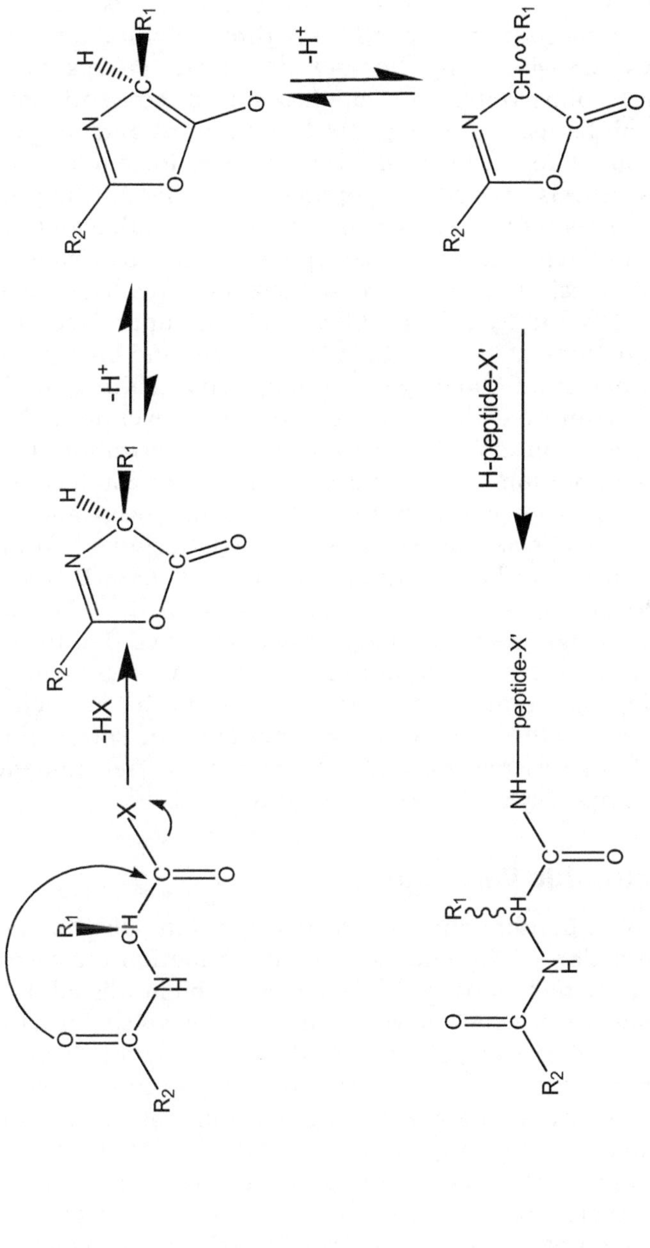

Scheme 5.5 Epimerization *via* oxazolone formation.[165]

pre-activation time, base-free activation,[168] change of base from *N,N*-diisopropylethylamine (DIEA) to 2,4,6-trimethylpyridine (TMP)[166,169] or coupling with pre-formed 2,3,4,5,6-pentafluorophenyl (Pfp) esters.[170]

Higher temperatures have a positive effect on the degree of epimerization and hence it is an important parameter to be considered while employing microwave irradiation to SPPS. One of the methods for measuring the degree of epimerization in peptides is through chiral amino acid analysis using GC-MS with a chiral column.[156] Several groups have studied and evaluated epimerization on application of conventional heating methods like oil baths.[171] It was in 2007 that Collins and co-workers presented a systematic report on epimerization levels for microwave-assisted SPPS in the synthesis of a 20-*mer* peptide containing all the proteogenic amino acids.[38] They set the temperature to 80 °C for both coupling and N^{α}-deprotection and reported increased epimerization, particularly for the Cys, His and Asp residues, with increased temperatures. They also claimed that in case of His and Cys, the problem could be suppressed by lowering the temperature from 80 °C to 50 °C. In 2008, Kappe and his research group investigated a magainin-II analog containing a His residue as well as a Cys residue.[154] They inferred that SPPS assisted by conventional heating or microwave irradiation afforded comparable epimerization levels, and hence the incorporation of His and Cys should be conducted at room temperature to prevent epimerization.[154] This has been applied in the synthesis of β-amyloid which contains three His residues. Chiral GC-MS analysis showed a very low epimerization level at 0.3% D-His for room temperature coupling *versus* 7% for coupling at 86 °C.[172] Loffredo *et al.* advocated the change in solvent theory. They suggested that the change in solvent from DMF to the binary, aprotic mixture DMSO–toluene (1 : 3) can effectively suppress epimerization of most natural amino acids during microwave-assisted SPPS at 60 °C using either Fmoc or Boc chemistry. The concept failed for Cys residue and yet again it was recommended that heating was completely avoided or lowered to 50 °C.[155,172]

5.6.2 Aspartimide Formation

In the case of piperidine-induced N^{α}-deprotection (20% piperidine in DMF), epimerization of Asp and aspartimide formation increases drastically on raising the temperature.[38,173,174] This problem was addressed by a change to piperazine, which lowered the level of D-Asp formation from 9.6% to 1.2%, and the degree of aspartimide formation from 31.5% to 3.15%.[85] In contrast, several other groups have synthesized shorter peptides containing Asp where the N^{α}-deprotection was carried out under standard conditions (20% piperidine in DMF at 60 °C using microwave irradiation), resulting in only minor amounts of D-Asp present (0.7%), indicating that there are still some unsolved issues concerning heating during the N^{α}-deprotection step and it appears that the aspartimide formation is highly sequence dependent.[156]

5.7 Microwaves and Solid Supports: a Comparative Account

With the advent of microwave heating in peptide synthesis, traditional polystyrene (PS) resins are rapidly being replaced by the poly(ethylene glycol) (PEG) modified PS supports (TentaGel, TG)[175] and the fully PEG-based ChemMatrix (CM) resin.[176] This has been attributed to the fact that both the TG and the CM resins swell well in the common solvents used in peptide synthesis (DMF, NMP, DCM as well as TFA).[175,176] In order to compare the efficacy of different resins, Kappe and co-workers synthesized the nonapeptide H-GILTVSVAV-NH$_2$, with DIC/HOBt-mediated peptide couplings. They used 10 equivalents of reagents on different resins for coupling time of 20 min at 60 °C. From the series of experiments, they drew the following conclusions. (a) The Rink Amide MBHA PS resin (loading: 0.64 mmol g^{-1}) and the Rink Amide TG resin (loading: 0.24 mmol g^{-1}) both gave a crude yield of 85%, but in contrast to the MBHA PS resin, the amount of coupling reagents could be reduced to 5 equiv. without major changes in the crude peptide purity (83%); (b) using Rink Amide CM resin (loading: 0.50 mmol g^{-1}) and 10 equiv. or 5 equiv. of coupling reagents gave a crude purity of 90% and 91%; (c) at 75 °C, using only 3 equiv. of coupling reagents, the CM resin outperformed the TG resin, *i.e.* 71% *versus* 91% crude purity; (d) at 86 °C, the CM resin (3 equiv. of coupling reagents, 95% crude purity) also resulted in a higher crude purity of the nonapeptide sequence compared to the TG resin (5 equiv. of coupling reagents, 92% crude purity).[155] Galanis *et al.* seconded the superior performance of the PEG-based resins by demonstrating an increase in crude purity of α-Conotoxin analogs on TG resin (loading: 0.24 mmol g^{-1}) compared to PS resin (loading: 0.43 mmol/g^{-1}).[176]

Apart from the commonly used PS, TG and CM resins, microwave-assisted SPPS has also been reportedly performed using PEGA, CLEAR and Wang resins.[177] Nishimura and co-workers established that PEGA (loading: 0.055 mmol g^{-1}) outperformed TG (loading: 0.26 mmol g^{-1}) in the synthesis of a MUC1-related glycopeptides of 20 amino acid residues. TG resin resulted in a crude peptide purity of 44%, as compared to 67% on using PEGA (poly(ethylene glycol)-poly-(*N,N*-dimethylacrylamide) copolymer) resin.[152] CLEAR resin was used for the synthesis of triple helical collagenmimetic lipopeptides in reasonable purity.[178] Papini and co-workers have reported the CSF114(Glc) peptide was synthesized in 98% crude purity using the pre-loaded Wang resin.[75]

The general impression that PEG-based resins outperform the PS-based resins, must be backed by the fact that choosing the optimal resin is an important parameter. The amount of inter- and intramolecular aggregation plays an important role in this scenario. Hence the degree of PEG, the amount of cross-linking and diversities in batches have significant influence on the probability of accessing the amino terminal of the growing peptidyl polymer.

5.8   Microwave-assisted Release of Peptides

It has been very well established by a number of groups that the use of microwave irradiation can effectively increase the speed of peptide release from the solid support and the concurrent side-chain deprotection.[58,59,178] These specific kinetic studies were conducted using standard resins and linker types (Rink Amide TG, Wang ChemMatrix, CLEAR Acid resin, Rink Amide PS, o-BAL PS).[8,59,178] It was reported that the cleavage time decreased from 2–5 hours down to minutes. Clearhout *et al.*[178] and Kluczyk *et al.,*[179] through extensive experiments illustrated that TFA under microwave irradiation can substitute HF in the cleavage of peptides from the Merrifield resin and meta-dialkoxy-BAL PS, respectively. Heating during TFA treatment may be of limited value for PEG-containing supports, as they may degrade at high temperatures.

Several researchers have developed some safety-catch type linkers, which release the peptide by nucleophilic displacement, *i.e.* under non-acidic conditions.[180–182] In these cases, microwave heating may indeed be very valuable. Park and Lee have taken advantage of microwave irradiation in the release of peptides from activated safety-catch linker by amines and showed that microwave heating increased the yield from 59% (25 °C, 100 min) to 92% (130 °C, 10 min) in the release of a small dipeptide.[180] Tofteng *et al.* studied the effect of microwave and conventional heating for the thiolytic release of protected peptide thioesters from the pyro-Glu linker.[181]

It can be concluded that, with the growing popularity of alternate and green sciences, microwave-assisted SPPS is gradually becoming one of the most sought-after synthesis techniques. Researchers, both academic and industrial, have successfully synthesized many peptides and some small protein molecules under microwave conditions. This success, however, does not necessarily ensure that dramatic rate- and yield-enhancements can always be achieved. Acylation of sterically demanding residues and syntheses of peptides that are prone to aggregation are the two categories of reactions which experienced dramatic results from microwave heating. Microwave instruments not only provide fast and precise heating and allow fast cooling by pressurized air, they also provide homogeneous and reproducible heating. The major challenges that need to be addressed include optimization of coupling time and temperatures. The activation of the amino acids should be performed as for conventional room temperature synthesis, except for the auxiliary nucleophile Oxyma, which is not stable at elevated temperature. Side reactions like epimerization, aspartimide formation and β-elimination need to be taken care of.

References

1. http://www.piercenet.com/method/peptide-synthesis.
2. www.thefreedictionary.com/Epitopes.
3. R. B. Merrifield, *J. Am. Chem. Soc.*, 1963, **85**, 2149.

4. L. A. Carpino, *J. Am. Chem. Soc.*, 1957, **79**, 4427.

5. F. C. McKay and N. F. Albertson, *J. Am. Chem. Soc.*, 1957, **79**, 4686.

6. G. W. Anderson and A. C. McGregor, *J. Am. Chem. Soc.*, 1957, **79**, 6180.

7. L. A. Carpino and G. Y. Han, *J. Org. Chem.*, 1972, **37**, 3404.

8. http://www-oc.chemie.uni-regensburg.de/OCP/ch/chv/oc22/script/006. pdf.

9. R. B. Merrifield, J. Singer and B. T. Chait, *Anal. Biochem.*, 1988, **174**, 399.

10. S. Mojsov, A. R. Mitchell and R. B. Merrifield, *J. Org. Chem.*, 2002, **45**, 555.

11. W. Konig and R. Geiger, *Chem. Ber*, 1970, **103**, 788.

12. L. A. Carpino, A. Elfaham, C. A. Minor and F. Albericio, *Chem. Commun.*, 1994, **24**, 201.

13. S.-Y. Han and Y.-A. Kim, *Tetrahedron*, 2004, **60**, 2447.

14. L. A. Carpino, *J. Am. Chem. Soc.*, 1993, **115**, 4397.

15. V. Dourtoglou, B. Gross, V. Lambropoulou and C. Zioudrou, *Synthesis*, 1984, **07**, 572.

16. L. A. Carpino, H. Imazumi, A. El-Faham, F. J. Ferrer, C. Zhang, Y. Lee, B. M. Foxman, P. Henklein, C. Hanay, C. Mugge, H. Wenschuh, J. Klose, M. Beyermann and M. Bienert, *Angew.Chem., Int. Ed.*, 2002, **41**, 441.

17. J. Coste, D. Lenguyen and B. Castro, *Tetrahedron Lett.*, 1990, **31**, 205.

18. A. El-Faham, R. S. Funosas, R. Prohens and F. Albericio, *Chem. – Eur. J.*, 2009, **15**, 9404.

19. A. Hyde, T. Johnson, D. Owen, M. Quibell and R. C. Sheppard, *Int. J. Pept. Protein Res.*, 1994, **43**, 431.

20. M. Dettin, S. Pegoraro, P. Rovero, S. Bicciato, A. Bagno and C. DiBello, *J. Pept. Res.*, 1997, **49**, 103.

21. B. D. Larsen and A. Holm, *J. Pept. Res.*, 1998, **52**, 470.

22. B. Henkel and E. Bayer, *J. Pept. Sci.*, 1998, **4**, 461.

23. J. F. McNamara, H. Lombardo, S. K. Pillai, I. Jensen, F. Albericio and S. A. Kates, *J. Pept. Sci.*, 2000, **6**, 512.

24. L. A. Carpino, E. Krause, C. D. Sferdean, M. Schumann, H. Fabian, M. Bienert and M. Beyermann, *Tetrahedron Lett.*, 2004, **45**, 7519.

25. T. Wohr, F. Wahl, A. Nefzi, B. Rohwedder, T. Sato, X. Sun and M. Mutter, *J. Am. Chem. Soc.*, 1996, **118**, 9218.

26. D. Seebach, A. Thaler and A. K. Beck, *Helv. Chim. Acta*, 1989, **72**, 857.

27. A. Thaler, D. Seebach and F. Cardinaux, *Helv. Chim. Acta*, 1991, **74**, 628.

28. K. Barlos, D. Papaioannou, S. Patrianakou and T. Tsegenidis, *Liebigs Ann. Chem.*, 1986, **11**, 1950.

29. B. E. Kaplan, L. J. Hefta, R. C. Blake, K. M. Swiderek and J. E. Shively, *J. Pept. Res.*, 1998, **52**, 249.

30. A. K. Rabinovich and J. E. Rivier, *Am. Biotechnol. Lab*, 1994, **12**, 48.

31. J. P. Tam, *Int. J. Pept. Protein Res.*, 1987, **29**, 421.

32. L. M. Varanda and M. T. M. Miranda, *J. Pept. Res.*, 1997, **50**, 102.

33. H. M. Yu, S. T. Chen and K. T. Wang, *J. Org. Chem.*, 1992, **57**, 4781.

34. S. Wang and G. L. Foutch, *Chem. Eng. Sci.*, 1991, **46**, 2373.

35. M. Erdelyi and A. Gogoll, *Synthesis*, 2002, 1592.

36. S. L. Pedersen, A. P. Tofteng, L. Malik and K. J. Jensen, *Chem. Soc. Rev.*, 2012, **41**, 1826.

37. B. Bacsa, K. Horváti, S. Bõsze, F. Andreae and C. O. Kappe, *J. Org. Chem.*, 2008, **73**, 7532.

38. S. A. Palasek, Z. J. Cox and J. M. Collins, *J. Pept. Sci.*, 2007, **13**, 143.

39. J. M. Collins and N. E. Leadbeater, *Org. Biomol. Chem.*, 2007, **5**, 1141.

40. H. Rodríguez, M. Suarez and F. Albericio, *J. Pept. Sci.*, 2010, **16**, 136.

41. R. Subirós-Funosas, R. Prohens, R. Barbas, A. El-Faham and F. Albericio, *Chem. – Eur. J*, 2009, **15**, 9394.

42. L. Malik, A. P. Tofteng, S. L. Pedersen, K. K. Sorensen and K. J. Jensen, *J. Pept. Sci.*, 2010, **16**, 506.

43. A. P. Tofteng, L. Malik, S. L. Pedersen, K. K. Sorensen and K. J. Jensen, *Chem. Today*, 2011, **29**, 28.

44. S. Härterich, S. Koschatzky, J. Einsiedel and P. Gmeiner, *Bioorg. Med. Chem.*, 2008, **16**, 9359.

45. O. Prante, J. Einsiedel, R. Haubner, P. Gmeiner, H.-J. Wester, T. Kuwert and S. Maschauer, *Bioconjugate Chem.*, 2007, **18**, 254.

46. S. Maschauer, J. Einsiedel, C. Hocke, H. Hübner, T. Kuwert, P. Gmeiner and O. Prante, *Med. Chem. Lett*, 2010, **1**, 224.

47. S. Lochner, J. Einsiedel, G. Schaefer, C. Berens, W. Hillen and P. Gmeiner, *Bioorg. Med. Chem.*, 2010, **18**, 6127.

48. D. Singer, T. Zauner, M. Genz, R. Hoffmann and T. Zuchner, *J. Pept. Sci.*, 2010, **16**, 358.

49. B. Bacsa, S. Bösze and C. O. Kappe, *J. Org. Chem.*, 2010, **75**, 2103.

50. M. P. D. Hatfield, R. F. Murphy and S. Lovas, *Biopolymers*, 2010, **93**, 442.

51. P. G. Katsoyannis, *Science*, 1966, **154**, 1509.

52. J. P. Mayer, F. Zhang and R. D. DiMarchi, *Pept. Sci.*, 2007, **88**, 687.

53. A. P. Tofteng, K. J. Jensen, L. Schaffer and T. Hoeg-Jensen, *ChemBioChem*, 2008, **9**, 2989.

54. F. Nissen, T. E. Kraft, T. Ruppert, M. Eisenhut, U. Haberkorn and W. Mier, *Tetrahedron Lett.*, 2010, **51**, 6216.

55. B. Indrevoll, R. Bjerke, D. Mantzilas, E. Hvattum and A. Rogstad, *Proceedings of the 31st European Peptide Symposium*, Copenhagen, Denmark, 2010, 62.

56. T. Attard, N. O'Brien-Simpson and E. Reynolds, *Int. J. Pept. Res. Ther.*, 2007, **13**, 447.

57. J. W. Perich, N. J. Ede, S. Eagle and A. M. Bray, *Lett. Pept. Sci.*, 1999, **6**, 91.

58. P. Harris, G. Williams, P. Shepherd and M. Brimble, *Int. J. Pept. Res. Ther.*, 2008, **14**, 387.

59. M. Brandt, S. Gammeltoft and K. Jensen, *Int. J. Pept. Res. Ther.*, 2006, **12**, 349.

60. J. Lukszo, D. Patterson, F. Albericio and S. A. Kates, *Lett. Pept. Sci.*, 1996, **3**, 157.

61. T. Attard, N. O'Brien-Simpson and E. Reynolds, *Int. J. Pept. Res. Ther.*, 2009, **15**, 69.

62. http://advancedpeptides.com/glycopeptides/.
63. H. Paulsen, *Angew. Chem., Int. Ed.*, 1990, **29**, 823.
64. H. Kunz, *Pure Appl. Chem.*, 1993, **65**, 1223.
65. M. Meldal, *Curr. Opin. Struct. Biol.*, 1994, **4**, 710.
66. A. El-Faham and F. Albericio, *Chem. Rev.*, 2011, **111**, 6557.
67. M. R. Pratt and C. R. Bertozzi, *Chem. Soc. Rev.*, 2005, **34**, 58.
68. T. Becker, S. Dziadek, S. Wittrock and H. Kunz, *Curr. Cancer Drug Targets*, 2006, **6**, 491.
69. T. Matsushita, H. Hinou, M. Fumoto, M. Kurogochi, N. Fujitani, H. Shimizu and S. I. Nishimura, *J. Org. Chem.*, 2006, **71**, 3051.
70. T. Matsushita, H. Hinou, M. Kurogochi, H. Shimizu and S.-I. Nishimura, *Org. Lett.*, 2005, **7**, 877.
71. M. Fumoto, H. Hinou, T. Ohta, T. Ito, K. Yamada, A. Takimoto, H. Kondo, H. Shimizu, T. Inazu, Y. Nakahara and S. I. Nishimura, *J. Am. Chem. Soc.*, 2005, **127**, 11804.
72. K. Naruchi, T. Hamamoto, M. Kurogochi, H. Hinou, H. Shimizu, T. Matsushita, N. Fujitani, H. Kondo and S.-I. Nishimura, *J. Org. Chem.*, 2006, **71**, 9609.
73. T. Matsushita, R. Sadamoto, N. Ohyabu, H. Nakata, M. Fumoto, N. Fujitani, Y. Takegawa, T. Sakamoto, M. Kurogochi, H. Hinou, H. Shimizu, T. Ito, K. Naruchi, H. Togame, H. Takemoto, H. Kondo and S.-I. Nishimura, *Biochemistry*, 2009, **48**, 11117.
74. N. Ohyabu, H. Hinou, T. Matsushita, R. Izumi, H. Shimizu, K. Kawamoto, Y. Numata, H. Togame, H. Takemoto, H. Kondo and S.-I. Nishimura, *J. Am. Chem. Soc.*, 2009, **131**, 17102.
75. F. Rizzolo, G. Sabatino, M. Chelli, P. Rovero and A. Papini, *Int. J. Pept. Res. Ther.*, 2007, **13**, 203.
76. S. Pandey, E. Peroni, M. C. Alcaro, F. Rizzolo, M. Chelli, P. Rovero, F. Lolli and A. M. Papini, *Proceedings of the 31st European Peptide Symposium*, Copenhagen, Denmark, 2010, 4.
77. A. Carotenuto, M. C. Alcaro, M. R. Saviello, E. Peroni, F. Nuti, A. M. Papini, E. Novellino and P. Rovero, *J. Med. Chem.*, 2008, **51**, 5304.
78. A. Heggemann, C. Budke, B. Schomburg, Z. Majer, M. Wisbrock, T. Koop and N. Sewald, *Amino Acids*, 2010, **38**, 213.
79. N. Miller, G. M. Williams and M. A. Brimble, *Org. Lett.*, 2009, **11**, 2409.
80. L. Nagel, C. Plattner, C. Budke, Z. Majer, T. Koop and N. Sewald, *Proceedings of the 31st European Peptide Symposium*, Copenhagen, Denmark, 2010, 86.
81. C. Heinlein, D. Varón Silva, A. Tröster, J. Schmidt, A. Gross and C. Unverzagt, *Angew. Chem., Int. Ed.*, 2011, **50**, 6406.
82. F. Haviv, T. D. Fitzpatrick, R. E. Swenson, C. J. Nichols, N. A. Mort, E. N. Bush, G. Diaz, G. Bammert and A. Nguyen, *J. Med. Chem.*, 1993, **36**, 363.
83. W. L. Cody, J. X. He, M. D. Reily, S. J. Haleen, D. M. Walker, E. L. Reyner, B. H. Stewart and A. M. Doherty, *J. Med. Chem.*, 1997, **40**, 2228.

84. M. A. Dechantsreiter, E. Planker, B. Matha, E. Lohof, G. Holzemann, A. Jonczyk, S. L. Goodman and H. Kessler, *J. Med. Chem.*, 1999, **42**, 3033.

85. J. Samanen, F. Ali, T. Romoff, R. Calvo, E. Sorenson, J. Vasko, B. Storer, D. Berry and D. Bennett, *J. Med. Chem.*, 1991, **34**, 3114.

86. P. L. Barker, S. Bullens, S. Bunting, D. J. Burdick, K. S. Chan, T. Deisher, C. Eigenbrot, T. R. Gadek and R. Gantzos, *J. Med. Chem.*, 1992, **35**, 2040; P. Wipf, *Chem. Rev.*, 1995, **95**, 2115.

87. P. Wipf, *Chem. Rev.*, 1995, **95**, 2115.

88. J. Chatterjee, F. Rechenmacher and H. Kesseler, *Angew. Chem., Int. Ed.*, 2013, **52**, 254.

89. E. Biron, J. Chatterjee and H. Kessler, *J. Pept. Sci.*, 2006, **12**, 213.

90. J. Chatterjee, C. Gilon, A. Hoffman and H. Kessler, *Acc. Chem. Res.*, 2008, **41**, 1331.

91. S. C. Miller and T. S. Scanlan, *J. Am. Chem. Soc.*, 1997, **119**, 2301.

92. R. D. Tung and D. H. Rich, *J. Am. Chem. Soc.*, 1985, **107**, 4342.

93. M. Teixido, F. Albericio and E. Giralt, *J. Pept. Res.*, 2005, **65**, 153.

94. Y. M. Angell, C. Garcia-Echeverria and D. H. Rich, *Tetrahedron Lett.*, 1994, **35**, 5981.

95. L. A. Carpino, A. Elfaham and F. Albericio, *J. Org. Chem.*, 1995, **60**, 3561.

96. E. Falb, T. Yechezkel, Y. Salitra and C. Gilon, *J. Pept. Res.*, 1999, **53**, 507.

97. H. Rodríguez, M. Suarez and F. Albericio, *J. Pept. Sci.*, 2010, **16**, 136.

98. M. Werder, H. Hauser, S. Abele and D. Seebach, *Helv. Chim. Acta*, 1999, **82**, 1774.

99. Y. Hamuro, J. P. Schneider and W. F. DeGrado, *J. Am. Chem. Soc.*, 1999, **121**, 12200; E. A. Porter, X. Wang, H.-S. Lee, B. Weisblum and S. H. Gellman, *Nature*, 2000, **404**, 565.

100. A. J. Karlsson, W. C. Pomerantz, B. Weisblum, S. H. Gellman and S. P. Palecek, *J. Am. Chem. Soc.*, 2006, **128**, 12630.

101. A. D. Bautista, J. S. Appelbaum, C. J. Craig, J. Michel and A. Schepartz, *J. Am. Chem. Soc.*, 2010, **132**, 2904.

102. E. A. Harker, D. S. Daniels, D. A. Guarracino and A. Schepartz, *Bioorg. Med. Chem.*, 2009, **17**, 2038.

103. J. K. Murray, J. D. Sadowsky, M. Scalf, L. M. Smith, Y. Tomita and S. H. Gellman, *J. Comb. Chem.*, 2008, **10**, 204.

104. A. D. Bautista, O. M. Stephens, L. Wang, R. A. Domaoal, K. S. Anderson and A. Schepartz, *Bioorg. Med. Chem. Lett.*, 2009, **19**, 3736.

105. J. K. Murray and S. H. Gellman, *Org. Lett.*, 2005, **7**, 1517.

106. J. K. Murray and S. H. Gellman, *J. Comb. Chem.*, 2006, **8**, 58.

107. J. K. Murray and S. H. Gellman, *Nat. Protoc.*, 2007, **2**, 624.

108. W. C. Pomerantz, N. L. Abbott and S. H. Gellman, *J. Am. Chem. Soc.*, 2006, **128**, 8730.

109. H. K. Cui, Y. Guo, Y. He, F. L. Wang, H. N. Chang, Y. J. Wang, F. M. Wu, C. L. Tian and L. Liu, *Angew. Chem.*, 2013, **125**, 9737.

110. W. C. Pomerantz, K. D. Cadwell, Y.-J. Hsu, S. H. Gellman and N. L. Abbott, *Chem. Mater.*, 2007, **19**, 4436.

111. C. Acevedo-Vélez, G. Andre, Y. F. Dufrêne, S. H. Gellman and N. L. Abbott, *J. Am. Chem. Soc.*, 2011, **133**, 3981.
112. E. J. Petersson and A. Schepartz, *J. Am. Chem. Soc.*, 2008, **130**, 821.
113. S. A. Fowler and H. E. Blackwell, *Org. Biomol. Chem.*, 2009, 7, 1508.
114. R. N. Zuckermann and T. Kodadek, *Curr. Opin. Mol. Ther.*, 2009, **11**, 299.
115. N. J. Brown, J. Johansson and A. E. Barron, *Acc. Chem. Res.*, 2008, **41**, 1409.
116. J. T. Nguyen, C. W. Turck, F. E. Cohen, R. N. Zuckermann and W. A. Lim, *Science*, 1998, **282**, 2088.
117. D. G. Udugamasooriya, S. P. Dineen, R. A. Brekken and T. Kodadek, *J. Am. Chem. Soc.*, 2008, **130**, 5744.
118. M. M. Reddy, R. Wilson, J. Wilson, S. Connell, A. Gocke, L. Hynan, D. German and T. Kodadek, *Cell*, 2011, **144**, 132.
119. K. T. Nam, S. A. Shelby, P. H. Choi, A. B. Marciel, R. Chen, L. Tan, T. K. Chu, R. A. Mesch, B. C. Lee, M. D. Connolly, C. Kisielowski and R. N. Zuckermann, *Nat. Mater.*, 2010, **9**, 464.
120. R. N. Zuckermann, E. J. Martin, D. C. Spellmeyer, G. B. Stauber, K. R. Shoemaker, J. M. Kerr, G. M. Figliozzi, D. A. Goff, M. A. Siani, R. J. Simon *et al.*, *J. Med. Chem.*, 1994, **37**, 2678.
121. R. N. Zuckermann, J. M. Kerr, S. B. H. Kent and W. H. Moos, *J. Am. Chem. Soc.*, 1992, **114**, 10646.
122. J. Seo, B. C. Lee and R. N. Zuckermann, *Compr. Biomater.*, 2011, **2**, 53.
123. G. M. Figliozzi, R. Goldsmith, S. C. Ng, S. C. Banville, R. N. Zuckermann and N. A. John, *Methods Enzymol.*, 1996, **267**, 437.
124. P. G. Alluri, M. M. Reddy, K. Bachhawat-Sikder, H. J. Olivos and T. Kodadek, *J. Am. Chem. Soc.*, 2003, **125**, 13995.
125. H. J. Olivos, P. G. Alluri, M. M. Reddy, D. Salony and T. Kodadek, *Org. Lett.*, 2002, **4**, 4057.
126. B. C. Gorske and H. E. Blackwell, *J. Am. Chem. Soc.*, 2006, **128**, 14378.
127. B. C. Gorske, S. A. Jewell, E. J. Guerard and H. E. Blackwell, *Org. Lett.*, 2005, 7, 1521.
128. J. Messeguer, N. Cortés, N. García-Sanz, G. Navarro-Vendrell, A. Ferrer-Montiel and A. Messeguer, *J. Comb. Chem.*, 2008, **10**, 974.
129. B. C. Gorske, J. R. Stringer, B. L. Bastian, S. A. Fowler and H. E. Blackwell, *J. Am. Chem. Soc.*, 2009, **131**, 16555.
130. G. D. Sala, B. Nardone, F. De Riccardis and I. Izzo, *Org. Biomol. Chem.*, 2013, **11**, 726.
131. I. Peretto, R. M. Sanchez-Martin, X.-h. Wang, J. Ellard, S. Mittoo and M. Bradley, *Chem. Commun.*, 2003, **18**, 2312.
132. M. A. Fara, J. J. Díaz-Mochón and M. Bradley, *Tetrahedron Lett.*, 2006, **47**, 1011.
133. J. J. Diaz-Mochon, M. A. Fara, R. M. Sanchez-Martin and M. Bradley, *Tetrahedron Lett.*, 2008, **49**, 923.
134. B. Rudat, E. Birtalan, I. Thome, D. K. Kolmel, V. L. Horhoiu, M. D. Wissert, U. Lemmer, H. J. Eisler, T. S. Balaban and S. Brase, *J. Phys. Chem. B*, 2010, **114**, 13473.

135. A. F. Spatola and K. Darlak, *Tetrahedron*, 1988, **44**, 821.
136. Y. Sasaki, W. A. Murphy, M. L. Heiman, V. A. Lance and D. H. Coy, *J. Med. Chem.*, 1987, **30**, 1162.
137. J. P. Meyer, P. Davis, K. B. Lee, F. Porreca, H. I. Yamamura and V. J. Hruby, *J. Med. Chem.*, 1995, **38**, 3462.
138. P. M. Hilaire, L. C. Alves, F. Herrera, M. Renil, S. J. Sanderson, J. C. Mottram, G. H. Coombs, M. A. Juliano, L. Juliano, J. Arevalo and M. Meldal, *J. Med. Chem.*, 2002, **45**, 1971.
139. V. Grand, A. Aubry, V. Dupont, A. Vicherat and M. Marraud, *J. Pept. Sci.*, 1996, **2**, 381.
140. D. H. Coy, S. J. Hocart and Y. Sasaki, *Tetrahedron*, 1988, **44**, 835.
141. V. Santagada, F. Frecentese, F. Fiorino, D. Cirillo, E. Perissutti, B. Severino, S. Terracciano and G. Caliendo, *QSAR Comb. Sci.*, 2004, **23**, 899.
142. M. S. Park, H. S. Oh, H. Cho and K. H. Lee, *Tetrahedron Lett.*, 2007, **48**, 1053.
143. B. P. Joshi, J.-w. Park, J.-m. Kim, C. R. Lohani, H. Cho and K.-H. Lee, *Tetrahedron Lett.*, 2008, **49**, 98.
144. H. N. Bramson, N. E. Thomas and E. T. Kaiser, *J. Biol. Chem.*, 1985, **260**, 15452.
145. G. S. Beligere and P. E. Dawson, *J. Am. Chem. Soc.*, 2000, **122**, 12079.
146. P. Silinski and M. C. Fitzgerald, *Biochemistry*, 2003, **42**, 6620.
147. L. Malik, A. P. Tofteng, S. L. Pedersen, K. K. Sorensen and K. J. Jensen, *J. Pept. Sci.*, 2010, **16**, 506.
148. Y. Tal-Gan, N. S. Freeman, S. Klein, A. Levitzki and C. Gilon, *Bioorg. Med. Chem.*, 2010, **18**, 2976.
149. M. P. D. Hatfield, R. F. Murphy and S. Lovas, *Biopolymers*, 2010, **93**, 442.
150. D. J. Lee, P. W. R. Harris, R. Kowalczyk, P. R. Dunbar and M. A. Brimble, *Synthesis*, 2010, **5**, 763.
151. A. J. Robinson, B. J. van Lierop, R. D. Garland, E. Teoh, J. Elaridi, J. P. Illesinghe and W. R. Jackson, *Chem. Commun.*, 2009, **28**, 4293.
152. T. Matsushita, H. Hinou, M. Fumoto, M. Kurogochi, N. Fujitani, H. Shimizu and S. I. Nishimura, *J. Org. Chem.*, 2006, **71**, 3051.
153. A. R. Katritzky, D. N. Haase, J. V. Johnson and A. Chung, *J. Org. Chem.*, 2009, **74**, 2028.
154. B. Bacsa, K. Horvati, S. Bosze, F. Andreae and C. O. Kappe, *J. Org. Chem.*, 2008, **73**, 7532.
155. C. Loffredo, N. A. Assuncão, J. Gerhardt and M. T. M. Miranda, *J. Pept. Sci.*, 2009, **15**, 808.
156. A. D. Bautista, J. S. Appelbaum, C. J. Craig, J. Michel and A. Schepartz, *J. Am. Chem. Soc.*, 2010, **132**, 2904.
157. J. K. Murray and S. H. Gellman, *Nature Protoc*, 2007, **2**, 624.
158. P. H. Chen, Y. H. Tseng, Y. Mou, Y. L. Tsai, S. M. Gua, S. J. Huang, S. S. F. Yu and J. C. Chan, *J. Am. Chem. Soc.*, 2008, **130**, 2862.
159. E. N. Murage, J. C. Schröeder, M. Beinborn and J. M. Ahn, *Bioorg. Med. Chem.*, 2008, **16**, 10106.

160. S. Harterich, S. Koschatzky, J. Einsiedel and P. Gmeiner, *Bioorg. Med. Chem.*, 2008, **16**, 9359.

161. O. Prante, J. Einsiedel, R. Haubner, P. Gmeiner, H.-J. Wester, T. Kuwert and S. Maschauer, *Bioconjugate Chem.*, 2007, **18**, 254.

162. S. Maschauer, J. Einsiedel, C. Hocke, H. Hubner, T. Kuwert, P. Gmeiner and O. Prante, *Med. Chem. Lett.*, 2010, **1**, 224.

163. S. Lochner, J. Einsiedel, G. Schaefer, C. Berens, W. Hillen and P. Gmeiner, *Bioorg. Med. Chem.*, 2010, **18**, 6127.

164. Y. Kajihara, N. Yamamoto, R. Okamoto, K. Hirano and T. Murase, *Chem. Rec.*, 2010, **10**, 80.

165. Y. Han, F. Albericio and G. Barany, *J. Org. Chem.*, 1997, **62**, 4307.

166. A. Di Fenza, M. Tancredi, C. Galoppini and P. Rovero, *Tetrahedron Lett.*, 1998, **39**, 8529.

167. T. Kaiser, G. J. Nichoson, H. J. Kohlbau and W. Voelter, *Tetrahedron Lett.*, 1996, **37**, 1187.

168. L. A. Carpino, D. Ionescu and A. El-Faham, *J. Org. Chem.*, 1996, **61**, 2460.

169. A. F. Spatola, K. Darlak and P. Romanovski, *Tetrahedron Lett.*, 1996, **37**, 591.

170. M. P. Souza, M. F. M. Tavares and M. T. M. Miranda, *Tetrahedron*, 2004, **60**, 4671.

171. B. Bacsa, S. Bõsze and C. O. Kappe, *J. Org. Chem.*, 2010, **75**, 2103.

172. C. Loffredo, N. A. Assuncao and M. T. M. Miranda, *Proceedings of the 20th American Peptide Symposium*, 2009, 165.

173. F. Nissen, T. E. Kraft, T. Ruppert, M. Eisenhut, U. Haberkorn and W. Mier, *Tetrahedron Lett.*, 2010, **51**, 6216.

174. P. Wright, D. Lloyd, W. Rapp and A. Andrus, *Tetrahedron Lett.*, 1993, **34**, 3373.

175. S. A. Camperi, M. M. Marani, N. B. Iannucci, S. Cote, F. Albericio and O. Cascone, *Tetrahedron Lett.*, 2005, **46**, 1561.

176. A. S. Galanis, F. Albericio and M. Grotli, *Pept. Sci.*, 2009, **92**, 23.

177. J. Banerjee, A. J. Hanson, W. W. Muhonen, J. B. Shabb and S. Mallik, *Nat. Protoc.*, 2010, **5**, 39.

178. S. Claerhout, T. Duchene, D. Tourwe and E. V. Van der Eycken, *Org. Biomol. Chem.*, 2010, **8**, 60.

179. A. Kluczyk, M. Rudowska, P. Stefanowicz and Z. Szewczuk, *J. Pept. Sci.*, 2010, **16**, 31.

180. J. W. Park and K. H. Lee, *Bull. Korean Chem. Soc.*, 2009, **30**, 2475.

181. A. P. Tofteng, K. K. Sorensen, K. W. Conde-Frieboes, T. Hoeg- Jensen and K. J. Jensen, *Angew. Chem., Int. Ed.*, 2009, **48**, 7411.

182. R. M. J. Liskamp, D. T. S. Rijkers, J. A. W. Kruijtzer and J. Kemmink, *ChemBioChem*, 2011, **12**, 1626.

CHAPTER 6

Hydrogels

6.1 Introduction

"Gels" and "hydrogels" are two words that can be used interchangeably to describe polymeric crosslinked network structures. The three-dimensional polymeric network of hydrogel is formed by the physical or chemical crosslinking of polymeric chains and has the ability to absorb and retain large volumes of water and aqueous solutions.[1] Chemically crosslinked hydrogels involve the covalent bond formation between functional groups present on polymer backbone with functional groups of crosslinking agents. In physically crosslinked hydrogels, the linkages formed between the polymer chains are hydrogen bonding, phase-separation, crystalline junctions *etc.* Physically developed hydrogels are devoid of any toxic chemicals but are less strong than chemically crosslinked hydrogels. The hydrogel structure is created by the hydrophilic groups or domains present in a polymeric network upon the hydration in an aqueous environment. Hydrogels have proven multifaceted applications. Hydrogels have been utilized by nature as mucus, vitreous humor, cartilage, tendons and blood clots.[2] The unique properties of natural and synthetic hydrogels make these materials the most appealing systems to be used for biomedical applications, such as drug delivery of soluble and insoluble drugs and biologically active molecules (*e.g.* proteins),[3,4] cell culture and tissue engineering.[5] The application of MWs in the synthesis of biomedical hydrogels is a relatively new research area and it is interesting to note that, before 2004, MW irradiation was used in the disinfection of hydrogel contact lenses.[6] Sosnik and co-workers have presented one of the most worthwhile chronologically compiled report on hydrogels prepared under microwave irradiation for biomedical applications in their latest review on microwave-assisted polymer synthesis as a tool in biomaterials science.[7] Due to their high water absorbing capacity, these materials are used for making baby nappies and adult incontinence pads,

RSC Green Chemistry No. 35
Microwave-Assisted Polymerization
By Anuradha Mishra, Tanvi Vats and James H. Clark
© Anuradha Mishra, Tanvi Vats and James H. Clark 2016
Published by the Royal Society of Chemistry, www.rsc.org

absorbent medical dressings, and controlled-release drug media. Hydrogels are extensively used for water remediation processes.[8–11] Edible gels are used widely in the food industry and mainly refer to gelling polysaccharides or hydrocolloids.[1]

Free-radical polymerization is the most common method to synthesize hydrogel systems. This reaction can be facilitated through the use of a large range of initiators and conditions. One method is through UV photo-polymerization, in which an initiator molecule is cleaved by the application of UV light. Another method is through thermal decomposition of the initiator molecule at elevated temperatures. Yet another synthetic route is through chemical initiation, where a chemical reaction causes the creation of a radical through the cleavage of a bond present in one of the reactant molecules. This initiation is through chemical means and does not require elevated temperatures.[12] As discussed in the previous chapters, although an extensive range of polymers has been synthesized using microwave (MW) irradiation,[13–16] this is usually undertaken using monomer reactants. The prime advantage of MW-assisted polymer synthesis is reduced reaction times and, in turn, minimized side reactions, several cases of addition polymerization have been reported that can occur without the need for an initiator.[17] The use of polymers as reactants in MW-assisted polymerizations is far less popular than the use of monomers. However, it is noteworthy to show that MW irradiation could serve as a valuable method of hydrogel synthesis by using a combination of polymeric reactants. The application of MW in the synthesis of biomedical hydrogels is a relatively new research area and it is interesting to note that, before 2004, MW irradiation was used in the disinfection of hydrogel contact lenses.[18] Besides the usual advantages, MW-assisted synthesis of sterile hydrogels is safer method. Though ionizing radiation finds application in polymers crosslinking to give sterile hydrogels,[19,20] the method requires a source of γ-rays or an electron beam, making the method more expensive, more hazardous and less common than MW reactors. Microwave irradiation is non-ionizing and can be considered safe.

Hydrogels may be prepared from monomers, pre-polymers and polymers. Copolymerization of monomers and low molecular weight prepolymers also results into hydrogel formation. Preparation methods of hydrogels may be classified on the basis of starting materials. This chapter deals with the MW-assisted preparation of polymers, co-polymers and polymer nanocomposites as hydrogels.

6.2 Homopolymeric Hydrogels

Homopolymeric hydrogels are composed of a polymer network derived from a single species of monomer.

In a recent study, Zhang and co-workers reported MW-assisted synthesis of biodegradable poly(2-hydroxyethylmethacrilate) (pHEMA) hydrogel for tissue engineering.[21] Though pHEMA hydrogels have been used in a plethora of biomedical activities[22] since 1960, biodegradability was a serious issue with them.[23] The authors addressed the issue by synthesizing biodegradable pHEMA hydrogel using MW-assisted polymerization of 2-hydroxyethyl methacrylate (HEMA). The reaction used potassium persulfate as the initiator, and polycaprolactone as the crosslinking agent. Characterization by FTIR and ^{1}HNMR confirmed the expected identity and purity of the hydrogel. Further, the degradation behavior of the hydrogel in PBS was determined by the weighing method and the cytotoxicity was measured by an MTT [3-(4,5-dimethylthiazol-2-yl)-2,5-diphenyltetrazolium bromide] assay, in order to evaluate the feasibility of using the hydrogel as a tissue engineering material. Biodegradation studies showed 75% uniform bulk degradation of the hydrogel over a period of seventeen days. In addition, the biodegradable hydrogel had no observable cytotoxicity toward L-929 fibroblast cells. MW irradiation was also evaluated in the condensation of poly(acrylic acid) (PAA) with amine-bearing adamantyl groups.[24,25] Ritter and co-workers described the MW-assisted synthesis of hydrogel-forming polymers based on PAA.[24] The authors carried out polymer analogous condensation reaction between adamantyl moieties bearing free amino groups and PAA by simply mixing both components and subsequent use of MW protocol. The reaction attained completion within 20 min of MW exposure without the addition of solvents or coupling agents (Scheme 6.1). The incorporation of hydrophobic pendant groups improved the hydrophobic interchain interactions in water, generating physical networks. Aqueous solutions of the sodium salts of the obtained hydrophobically modified PAAs showed a very high viscosity due to the intermolecular association of the hydrophobic side chains and the resulting formation of physical networks.

Jovanovic and co-workers studied the kinetics of isothermal formation of PAA hydrogels through polymerization of acrylic acid and crosslinking of the PAA.[26] A comparative account of conventionally heated reaction system and a MW heated reaction system was given in detail. The authors observed that in the MW heated system, the reaction rate constant of PAA hydrogel formation exhibited a mammoth increase between 32- to 43-fold, when compared with the conventionally heated system. An interesting analysis of the kinetics revealed that the isothermal kinetics of the PAA hydrogel formation during the microwave process could follow a first-order chemical reaction kinetics model, while a second-order kinetics was followed during the conventionally heated process. Besides the change in the rate order, the other kinetic parameters that changed in the MW heated system included the activation energy (E_a) which decreased by about 19% and the pre-exponential factor $(\ln A)$ which showed a decreased by 2.2 times. The decrease in activation energy, change in entropy of activation energy, and decrease in the pre-exponential value of PAA hydrogel formation under MW heating were caused with increased energy of the reactive species when compared with their energy in thermal activation. The authors related this increase in energy of the reactive species to the rapid transfer and absorption of the energy of microwave field to the existing reactive species.

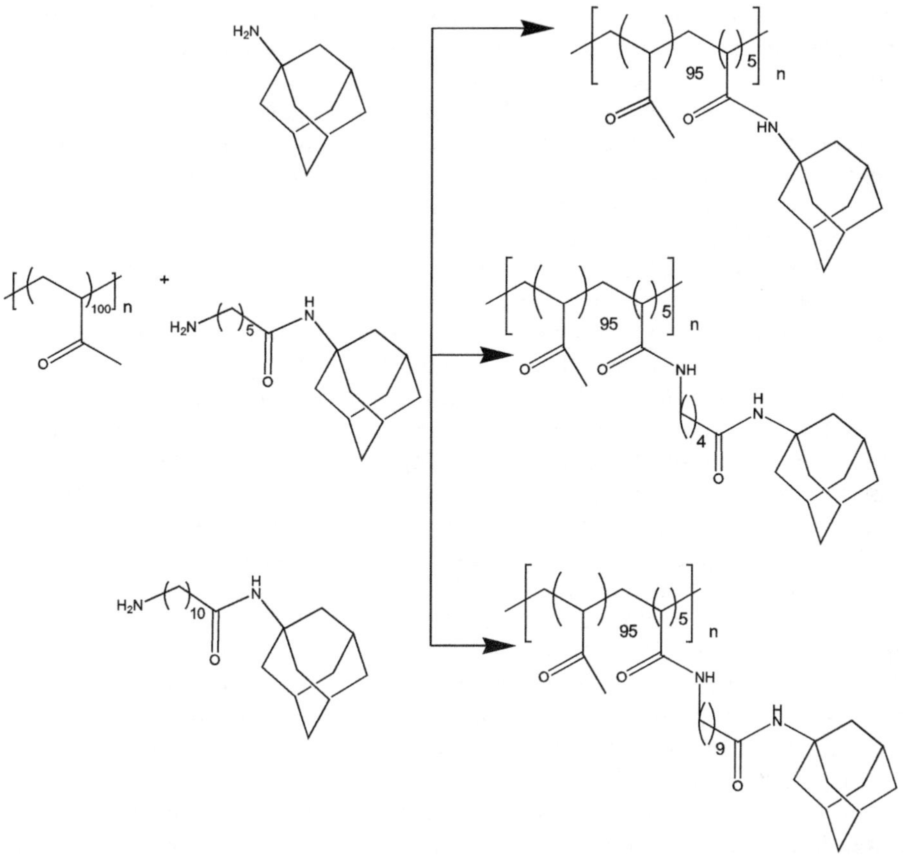

Scheme 6.1 Synthesis of hydrophobically modified poly(acrylic acids) by reaction of poly(acrylic acid) with amines with a reaction time of 20 mins under MW irradiation at 75 W and temperature around 250 °C.

Poly(acrylic acid) (PAA)-based controlled release carrier for fertilizers with ammonium nitrate or carbamide were prepared using microwave irradiation. The obtained hydrogels were characterized by swelling behavior and ammonium ion release. The swelling of hydrogels containing ammonium nitrate decreased with increased concentrations of ammonium nitrate. The reason is the presence of ionic groups during agrochemicals' release in water. These ions prevent water molecules from diffusion into the hydrogels, thus decreasing the swelling capacity of the hydrogels. The release rate is the highest during the first 24 hours, then reaches a plateau.[27]

Poly(*N*-isopropylacrylamide) (PNIPAM) derivatives are widely used as "smart" drug delivery matrices and tissue engineering scaffolds that can be injected by minimally invasive techniques. This is because of the fact that PNIPAM generates thermo-responsive physical hydrogels that are liquid at room temperature and gels at 37 °C; the temperature of the sol–gel transition is defined as the lower critical solution temperature and it is usually around

32 °C.[28] Shi and Liu synthesized a series of PNIPAM-based hydrogels under MW irradiation. The authors used poly(ethylene oxide)-600 (PEO-600) for a three-fold function; namely, reaction medium, microwave-absorbing agent and a pore-forming agent.[29] The MW irradiations gave an excellent yield of 98% within 1 min of the reaction time. The resultant PNIPAM hydrogels exhibited controllable properties such as pore size, equilibrium swelling ratios, and swelling/deswelling rates on changing the feed weight ratios of monomer (*N*-isopropylacrylamide, NIPAM) to PEO-600. These properties are well adapted to the different requirements for their potential application in many fields, such as biomedicine.

Zhao and co-workers presented a comparative account of the synthesis of crosslinked poly(*N*-isopropylacrylamide) (PNIPAM) by thermal and MW methods.[30] Their work involved the study of swelling and deswelling kinetics of PNIPAM hydrogels separately synthesized by means of MW irradiation and normal water-bath heating. As compared with the PN hydrogel synthesized by the conventional method, the PM hydrogel synthesized by MW irradiation had larger swelling and deswelling rate constants as well as lower swelling/deswelling activation energy due to its higher surface area and larger pore sizes, and demonstrated faster response behavior. The conventional thermal method conducted at 70, 80 and 90 °C for 24 h gave a yield of ~73%. In contrast, the MW method gave 87–100% yields in 5–30 min of the reaction time. With the aid of the SEM micrographs, the authors observed that MW-produced hydrogels showed a more porous structure that enabled more efficient water diffusion into or out of the network upon cooling/heating cycles.[30,31]

Another group of polymers that has benefited with the advent of microwaves is poly(2-oxazoline)s.[28,29] The emergence of laboratory MW reactors proved to be of great significance for the polymerization of 2-oxazolines. The synthetic route remarkably decreased reaction times with a factor of 60, maintaining the livingness of the polymerization.[32,33] Wiesbrock and co-workers did an extensive study on the MW-assisted synthesis of hydro-, amphi-, and lipogel libraries. They reported the influence of the ratio of poly(2-ethyl-2-oxazoline) *versus* poly(2-phenyl-2-oxazoline), the degree of crosslinking and the type of crosslinker on the swelling degree and the proton-mediated degradation of the gels. The study established the great potential of MW polymerized 2-oxazoline-based gels as toolbox for tailor-made hydro-, lipo- and amphigels.[34]

6.3 Copolymeric Hydrogels

Copolymeric hydrogels are made up of two or more different monomers with at least one hydrophilic component. Copolymeric hydrogels may be prepared by aqueous solutions containing appropriate combinations of polymers using microwave irradiation. With the aid of this novel method, it was possible to synthesize sterile hydrogels without the use of monomers, eliminating the need for the removal of unreacted species from the final product. MW irradiation of poly(vinyl alcohol) with either poly(acrylic acid) or poly(methylvinylether-alt-maleic anhydride), at 100–150 °C yielded hydrogels with large equilibrium swelling degrees of 500–1000 g/g. Material leached from both types of hydrogel show little cytotoxicity towards HT29 cells.[35] The polymerization technique involved several steps. The first step involved dissolution of AM and HEMA monomers and crosslinker in distilled water, followed by nitrogen bubbling and addition of the initiator. The polymerization was carried out at 300 W for 5 minutes in MW oven. Crosslinked hydrogels were obtained by adding glutaraldehyde and hydrochloric acid into the reaction mixture and continuing the polymerization for another 5 minutes.[35]

"Click chemistry" is a synthetic rationale, introduced approximately one decade ago, that relies on the quick and reliable synthesis of compounds by reacting small functional groups. Bardts and Ritter employed MWs for the first time to incorporate pendant cystamine[36] groups into poly(methacrylic acid) (PMAA) chains by the formation of amide bonds.[37] This thiol-ene "click" reaction was conducted at 80 W over 10 min. Then this derivative was reacted with allyl-derivatized PMAA by thiol-ene reactions to form crosslinked gels. The kinetics of this reaction were determined by ^{1}H NMR spectroscopy and the resulting hydrogels were analyzed by rheological measurements and differential scanning calorimetry (DSC). Surprisingly, MW irradiation was not evaluated in the crosslinking step and this reaction demanded 2 hours.

Kalaleh and co-workers presented a new one-pot synthesis method of poly(acrylate-coacrylamide) superabsorbent polymer *via* partial alkaline hydrolysis of acrylamide using microwave irradiation.[38] The method used potassium peroxodisulfate as an initiator and *N,N*-methylene bisacrylamide as crosslinker, and involved addition of adequate amounts of sodium hydroxide to a solution of acrylamide. After 15 minutes of stirring the mixture is exposed to MW irradiation at the power of 950 W. This method allowed hydrolysis, polymerization and gelation to take place in one pot during a very short reaction time of just 90 seconds. Apart from reducing the reaction time, the method eliminated the need to operate under inert atmosphere. Scanning electron microscopy (SEM) showed that the synthesized hydrogel has a macroporous structure.

Gibson *et al.* pioneered in the implementation of MW irradiation for the synthesis of crosslinked hydrogels.[25] The authors presented facile synthesis by the microwave-assisted route to achieve fast conversions under solvent-free condition (Scheme 6.2). Poly(ethylene glycol)s of various molecular masses (Mn) 1000 to 8000 g/mol were reacted with methacrylic anhydride to form PEGDMs. A detailed and combined analysis of ^{1}H NMR and MALDI-TOF MS confirmed the formation of prepolymers of high purity and narrow mass distribution (PD < 1.02).

Crosslinked PEG and PEG-rich hydrogels have been extensively investigated for drug delivery and cell culture.[39–43] The notable feature of these reactions is that MW irradiation was involved only in the primary synthesis

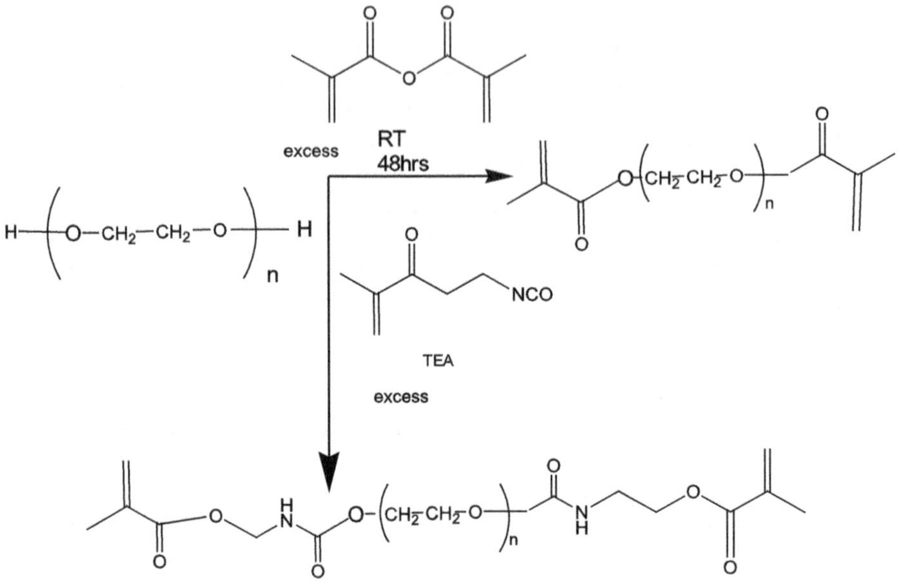

Scheme 6.2   Microwave-assisted synthesis of poly(ethylene glycol) dimethacrylates (PEGDM) and poly(ethylene glycol) urethane-dimethacrylates (PEGUDM).

of the photopolymerizable and biocompatible poly(ethylene glycol) dimethacrylate (PEGDMA) precursor.[40] A domestic oven with an operating power of 1100 W was used to obtain PEGDMA by reacting PEG of different molecular weights (1–8 kDa) with methacrylic anhydride under solvent-free conditions. Conversely, the conventional method involved the use of triethylamine. The MW-assisted reaction required just 10 min in contrast to 4 days in conventional thermal method. This drastic difference in reaction time could be attributed to the high MW-absorbing capacity of PEG.[44]

Meena *et al.* synthesized adhesive by graft copolymerizing κ-carrageenan and polyacrylamide copolymers with different N contents.[45] Pandey *et al.* evaluated the effect of solubilized and dispersed bacterial cellulose (BC) on the physicochemical characteristics and drug release profile of hydrogels synthesized using biopolymers.[46] The authors synthesized superabsorbent hydrogels by graft polymerization of acrylamide (AM) on BC solubilized in an NaOH/urea solvent system and *N*,′-methylenebisacrylamide as a crosslinker under MW irradiation at 340 W for 30 s. To synthesize the BC/PAM hydrogels, AM was added to freshly prepared BC solution, followed by the addition of potassium persulfate as an initiator and *N*,′-methylenebisacrylamide as a crosslinker to obtain hydrogels. The hydrogels exhibited pH and ionic responsive swelling behavior, with hydrogels prepared using solubilized BC having higher swelling ratios. Furthermore, compared to the hydrogels synthesized using dispersed BC, the hydrogels synthesized using solubilized BC showed higher porosity, drug loading efficiency and release. These results suggest the superiority of the hydrogels prepared using solubilized BC and that they should be explored further for oral drug delivery. Bhatia reported the MW-assisted synthesis of psyllium-*g*-poly(N-vinyl-2-pyrollidone).[47] Microwave-assisted graft co-polymerization of *N*-vinyl-2-pyrollidone on psyllium husk was optimized using central composite experimental design. The authors used ammonium persulphate as a redox initiator and *N*-vinyl pyrollidone as a crosslinking agent. It was observed that higher concentration of ammonium persulphate and lower concentration of *N*-vinyl pyrollidone increases the grafting efficiency. The authors also evaluated muco-adhesive properties of graft coplymerized psyllium gel using modified physical balance method and concluded that psyllium-*g*-(*N*-vinyl-2-pyrrolidone) can be posed as bioadhesive material for drug delivery.

6.4 Multipolymeric Hydrogels

Multipolymeric hydrogels are usually interpenetrating polymeric hydrogels (IPNs) or semi-IPNs. It is an important class of hydrogels. IPN is made of two independent crosslinked synthetic and/or natural polymer components, contained in a network form. In semi-IPN hydrogel, one component is a crosslinked polymer and other component is a non-crosslinked polymer.

Prasad and co-workers reported the synthesis of gel-forming polyvinylpyrrolidone (PVP)-grafted agar and κ-carragenan blends. This was perhaps the first reported MW-assisted synthesis of modified polysaccharides to be used in the biomedical field.[48] The reaction operated through MW induced formation of free radicals along the carbohydrate chain due to a localized overheating of –OH groups and the generation of O[•] reactive moieties which attack the reactive monomer, eventually initiating the polymerization. The MW-assisted reaction was completed within 2 min and because of their crosslinked nature, the products were not crystalline, as the pristine carbohydrates. It was reported that the dry copolymers displayed greater water-absorption capacity than the pristine derivatives. Though the water-holding capacities of the hydrogels increased, the mechanical strength decreased, and this made them a suitable material for absorbent wound dressings, tissue engineering matrices and topical drug delivery systems.[7]

Tanan and Saengsuwan reported MW-assisted synthesis of poly (acrylamide-*co*-2-hydroxyethyl methacrylate)/poly(vinyl alcohol) [(P(AM-*co*-HEMA)/PVA)] semi-IPN hydrogels.[49] These semi-IPN hydrogels were crosslinked by glutaraldehyde using ammonium persulfate as an initiator. The reaction involved the synthesis of [(P(AM-*co*-HEMA)] network in PVA aqueous solution, followed by glutaraldehyde crosslinking reagent, forming a semi-IPN structure. The authors gave a comparative account of the hydrogel synthesis by one-pot method and a two-step polymerization method.

Bajpai *et al.* investigated microwave-assisted synthesis of carboxymethyl psyllium and its development as semi-interpenetrating network with poly(acrylamide) for gastric delivery.[50] A wood pulp cellulose-based hydrogel material was prepared with poly(methyl vinyl ether-*co*-maleic acid), polyethylene glycol and softwood ECF kraft pulp *via* microwave esterification. The maximum water absorbency of the milled pulp fibers was 151 g/g, and it could retain a maximum of 67% of absorbed water after centrifugation at 770 rpm for 10 min. Infrared spectroscopy was used to confirm the esterification of poly(methyl vinyl ether-*co*-maleic acid) with the pulp cellulose.[51]

A microwave (MW)-assisted crosslinking process to prepare hydrogel-forming micro-needle (MN) arrays for *trans*-dermal drug delivery was evaluated. Conventionally, such MN arrays are prepared using processes that include a thermal crosslinking step. Polymeric MN arrays were prepared using poly(methyl vinyl ether-*alt*-maleic acid) crosslinked by reaction with poly(ethylene glycol) over 24 h at 80 °C. The analysis suggested that MN arrays prepared using the MW-assisted process had equivalent properties to those prepared conventionally but can be produced 30 times faster.[52]

In situ forming drug delivery systems composed of the thermoresponsive triblock copolymer PLGA-PEG-PLGA was studied as a dexamethasone delivery system. The system was prepared under microwave irradiation for five minutes.[53] A novel keratin-based polymer hydrogel was successfully prepared by graft-copolymerization of methacrylic acid onto keratin, crosslinked with *N,N'*-methylene-bis-acrylamide, then blended with agar. The material thus obtained may be used as a wound dressing or bandage.[54]

6.5   Polymer-nanocomposites as Hydrogels

With the advancement of nanoscience and technology, materials with multifold applications have come to light. Nanocomposite systems are one such material. The composites have been developed through the combination of synthetic polymers with a range of nanoparticulate fillers, like silica, metal and magnetic particles.[55] There are several ways the nanocomposites can be formed. The polymer can be used to coat the nanoparticles to form core-shell systems, the filler can be incorporated into small-scale polymer systems, such as liposomes and micelles, or the filler can be dispersed throughout a bulk polymer matrix.[56–58] The major advantage of these systems over the pure polymer is that the addition of the fillers can enhance properties (*e.g.*, mechanical properties) or can introduce new properties (*e.g.*, remote heating).[59] The composite systems are studied because of their ability to allow controlled response to an external stimulus (*i.e.*, electricity, magnetism, light, *etc.*); this response can be turned on and off and result in a rapid change in material properties.[60] These properties make composites highly researched materials, aimed at generating good commercial viability.

In a very interesting study, MW-assisted quantum dots synthesis was used to create nanohydrogels. Cadmium selenide/zinc sulfide (CdSe/ZnS) core-shell quantum dots coated with an amphiphilic polymer were incorporated in poly(NIPAM) (QD/PEI/poly(NIPAM)) hydrogels.[61] Luminescent semiconductor nanocrystals, or quantum dots possess unique optical and electrical characteristics due to quantum confinement effects.[62] High luminescence, near Gaussian narrow emission, broad absorption band and resistance to photobleaching are the meritorious properties which render quantum dots suitable for vast technological applications ranging from biomarkers,[63,64] light emitting diodes (LED's),[65,66] solar cells[67] and optical sensing.[68,69]

Common synthetic methods deployed for quantum dot synthesis are based on organometallic pyrolysis of precursors which often required heating at temperatures up to 300 °C. There were several disadvantages associated with the conventional heating methods employed; this affected the quality of the synthesized quantum dots. MW-assisted quantum dots synthesis enabled optimization of existing quantum dots synthesis methods. The major advantage of using a commercial MW oven is the precise control of the reaction parameters, faster reaction rates and synthesis of nearly monodisperse quantum dots samples with high photoluminescent (PL) intensity.

PNIPAM exhibits a temperature-dependent phase transition phenomenon and switches from being hydrophilic to hydrophobic and *vice versa*, in aqueous solution.[70] Thus, when optical properties of quantum dots are effectively combined with the physical properties of thermoresponsive polymers, they can serve as building blocks for optical nanothermometers.

In order to obtain a comprehensive understanding of the feasibility of developing quantum dots-thermoresponsive polymer nanocomposite hydrogel based optical temperature sensors, the temperature-dependent emission spectra of the hydrogels were recorded as a function of the concentration of the incorporated quantum dots. These studies showed a definite relationship between the change in the position of the peak emission wavelength and concentration of the incorporated PEI coated quantum dots. The possibility of energy transfer between individual quantum dots was reduced due to the PEI coating and it was speculated that the changes in the physical environment *i.e.* the shrinking and swelling behavior of the poly(NIPAM) hydrogel due to the temperature change across the lower critical solution temperature caused the spectral properties of the quantum dots change accordingly. We can say that quantum dot-thermoresponsive polymer nanocomposite hydrogel-based optical temperature sensors is indeed a great research area and has a great potential for commercial application.

Nanohydrogels have been developed and studied with regard to their application in several biomedical fields also, *e.g.* separation techniques, soft-actuators and controlled drug delivery systems.[71] MW-assisted polysaccharide-based nanohydrogels stand as a testimony to the acceptance of polysaccharide-based semisynthetic polymers as eco-friendly alternatives in several fields. Polyacrylic acid, guar gum grafted PAA/cloisite superabsorbent nanocomposites were prepared by both conventional and MW methods.[72] Chitosan hydrogel has been prepared with a variety of different shapes, geometries, and formulations that include liquid gels, powders, beads, films, tablets, capsules, microspheres, micro particles, sponges, nanofibrils, textile fibers and inorganic composites.[73] Chitosan has exhibited properties to facilitate wound healing, reduce serum cholesterol levels and to stimulate the immune system. Drugs can be easily encapsulated into the nanohydrogels, which are stable enough to circulate in the blood stream over long periods without allowing much leaching.[74] However, the nanoparticles easily degrade once inside the target cells, are hydrolyzed by enzymes in the cell cytoplasm, and release the drug being carried. Last, but by no means least, the fragments of the degraded nanohydrogels can be eliminated by the body and show no toxic effects on cells.[75]

Devi and co-workers, presented a report on MW-assisted synthesis of nanoparticle composites for drug delivery systems.[76] They synthesized the carrier by block polymerization of acryl amide monomer and poly (ethylene glycol) onto chitosan using glutaraldehyde as a crosslinker. The authors performed a comparative synthesis of chitosan using both conventional heating and MW irradiation (in a domestic MW oven at 480 W). The conventional heating took about seven hours, while the MW-assisted reaction comprised repeated cycles of two minutes of heating and one minute of cooling for a total of 50 minutes. Silver and gold nanoparticles were loaded *in vitro* on the hydrogels and the release behavior of nanohydrogels for the drug ampicillin was studied. The report concluded that the nanohydrogels released 51% of the drug at the end of 7 hours, thus indicating the sustained drug release, which is crucial to increase the drug bioavailability and prolong the therapeutic effect.[76] Novel hydrogels containing nanosilver for biomedical applications have been synthesized by a unique method, which involves formation of silver nanoparticles within hydrogels under microwave irradiation. Nanoparticles, with a diameter of about 10 nm, were anchored onto the polymer network.[77] Novel nanocomposite hydrogels composed of Noveon® AA-1 Polycarbophil and acrylamide and gold/silver nanoparticles were developed. Inorganic nanoparticles were produced by the nucleation of Ag^+ and Au^{3+} salts with extracts of mint leaf that formed within the hydrogel system. These nanocomposite hydrogels have shown good antibacterial property.[78]

Creating inorganic nanowire hydrogels/aerogels using various materials and inexpensive means remains an outstanding challenge despite their importance for many applications. Production of highly porous inorganic nanowire hydrogel/aerogel on a large scale and at low cost has been achieved by *in situ* hydrothermal synthesis of one-dimensional (1D) nanowires, directly forming a crosslinking network during the synthesis process. Such a method not only offers great simplicity but also allows the interconnecting nanowires to have much longer length, remarkable porosity, surface area and extremely low densities. These hydrogels/aerogels were mechanically robust and can have superelasticity by tuning the synthesis conditions. The nanowires in the hydrogels/aerogels serve both as structural support and active sites, for example, for catalysis or absorption. The report indicated that this method for nanowire hydrogels/aerogels production was very economical and the products have a great potential to be used in catalysis, sensing, adsorption, energy storage and beyond.[79]

Overall, the data converge to crown MW as the fastest and most efficient method for the production of water-containing matrices. Nevertheless, these incipient studies still appear as relatively isolated efforts and thus, the potential of this area still remains unexploited. The expanse of the area covered by the hydrogel application will definitely increase by the inculcation of MW-assisted synthesis.

References

1. S. K. H. Gulrez, S. Al-Assaf and G. O Phillips, *Progress in Molecular and Environmental Bioengineering-From Analysis and Modeling to Technology Applications*, ed. A. Carpi, 2011, ISBN 978-953-307-268-5, InTech, DOI: 10.5772/24553.
2. N. Das, *Int. J. Pharm. Pharm. Sci.*, 2013, **5**, 55.
3. C. C. Lin and A. T. Metters, *Adv. Drug Delivery*, 2006, **58**, 1379.
4. J. K. Oh, R. Drumright, D. J. Siegwart and K. Matyjaszewski, *Prog. Polym. Sci.*, 2008, **33**, 448.
5. K. Y. Lee and D. J. Mooney, *Chem*, 2001, **101**, 1869.
6. A. Crabbe and P. Thompson, *Optom. Vis. Sci.*, 2004, **81**, 471.
7. A. Sosnik, G. Gotelli and G. A. Abraham, *Prog. Polym. Sci.*, 2011, **36**, 1050.
8. J. L. Drury and D. J. Mooney, *Biomaterials*, 2003, **24**, 4337.
9. S. Van Vlierberghe, P. Dubruel and E. Schacht, *Biomacromolecules*, 2011, **12**, 1387.
10. J. S. Boateng, K. H. Matthews, H. N. E. Stevens and G. M. Eccleston, *J. Pharm. Sci.*, 2008, **97**, 2892.
11. Y. Qiu and K. Park, *Adv. Drug Delivery Rev.*, 2001, **53**, 321.
12. A. S. Sarac, *Prog. Polym. Sci.*, 1999, **24**, 1149.
13. A. K. A. Silva, C. Richard, M. Bessodes, D. Scherman and O. W. Merten, *Biomacromolecules*, 2009, **1**, 9.
14. F. Wiesbrock, R. Hoogenboom and U. S. Schubert, *Macromol. Rapid Commun.*, 2004, **25**, 1739.
15. R. Hoogenboom and U. S. Schubert, *Macromol. Rapid Commun.*, 2007, **28**, 368.
16. C. Ebner, T. Bodner, F. Stelzer and F. Wiesbrock, *Macromol. Rapid Commun.*, 2011, **32**, 254.
17. M. Teffal and A. Gourdenne, *Eur. Polym. J.*, 1983, **19**, 543.
18. A. Crabbe and P. Thompson, *Optom. Vis. Sci.*, 2004, **81**, 471.
19. J. M. Rosiak and P. Ulanski, *Radiat. Phys. Chem.*, 1999, **55**, 139.
20. Z. S. Nurkeeva, G. A. Mun, A. V. Dubolazov and V. V. Khutoryanskiy, *Macromol. Biosci.*, 2005, **5**, 424.
21. L. Zhang, G. J. Zheng, Y. T. Guo, L. Zhou, J. Du and H. He, *Asian Pac. J. Trop. Med.*, 2014, **7**, 136.
22. S. Atzet, S. Curtin, P. Trinh, S. Bryant and B. Ratne, *Biomacromolecules*, 2008, **9**, 3370.
23. K. S. Anseth and C. N. Bowman, *J. Phys. Chem. B*, 2001, **105**, 8069.

24. O. Kretschmann, S. Schmitz and H. Ritter, *Macromol. Rapid Commun.*, 2007, **28**, 1265.
25. S. Lin-Gibson, S. A. Bencherif, J. A. Cooper, S. J. Wetzel, J. M. Antonucci, B. M. Vogel, F. Horkay and N. R. Washburn, *Biomacromolecules*, 2004, **5**, 1280.
26. J. Jovanovic and B. Adnadjevic, *J. Appl. Polym. Sci.*, 2010, **116**, 55.
27. B. Tyliszczak, J. Polaczek and K. Pielichowski, *Pol. J. Environ. Stud.*, 2009, **18**, 475.
28. A. Sosnik, *ARS Pharm.*, 2007, **48**, 83.
29. S. Shi and L. Liu, *J. Appl. Polym. Sci.*, 2006, **102**, 4177.
30. Z. Zhao, Z. Li, Q. Xia, H. Xi and Y. Lin, *Eur. Polym. J.*, 2008, **44**, 1217.
31. Z. X. Zhao, Z. Li, Q. B. Xia, E. Bajalis, H. X. Xi and Y. S. Lin, *Chem. Eng. J*, 2008, **142**, 263.
32. Y. Chujo, K. Sada, K. Matsumoto and T. Saegusa, *Macromolecules*, 1990, **23**, 1234.
33. J. E. Puskas and Y. Chen, *Biomacromolecules*, 2004, **5**, 1141.
34. R. Hoogenboom, F. Wiesbrock, H. Y. Huang, M. A. M. Leenen, H. M. L. Thijs, S. F. G. M. van Nispen, M. V. Loop, C. A. Fustin, A. M. Jonas, J. F. Gohy and U. S. Schubert, *Macromolecules*, 2006, **39**, 4719.
35. J. P. Cook, G. W. Goodall, O. V. Khutoryanskaya and V. V Khutoryanskiy, *Macromol. Rapid Commun.*, 2012, **33**, 332.
36. M. Bardts and H. Ritter, *Macromol. Chem. Phys.*, 2010, **211**, 778.
37. H. C. Kolb, M. G. Finn and K. B. Sharpless, *Angew. Chem., Int. Ed.*, 2001, **40**, 2004.
38. http://arxiv.org/ftp/arxiv/papers/1502/1502.03639.pdf.
39. A. Sosnik, D. Cohn, J. San Roman and G. A. Abraham, *J. Biomater. Sci., Polym. Ed.*, 2003, **14**, 227.
40. A. Sosnik and D. Cohn, *Biomaterials*, 2004, **25**, 2851.
41. D. Cohn, A. Sosnik and S. Garty, *Biomacromolecules*, 2005, **6**, 1168.
42. A. Sosnik and M. V. Sefton, *Biomaterials*, 2005, **26**, 7425.
43. A. Sosnik, B. Leung, A. P. McGuigan and M. V. Sefton, *Tissue Eng.*, 2005, **11**, 1807.
44. A. H. V. Hove, B. D. Wilson and D. S. Benoit, *J. Visualized Exp.*, 2013, **80**, e50890, DOI: 10.3791/50890.
45. R. Meena, K. Prasad, G. Mehta and A. K. Siddhanta, *J. Appl. Polym. Sci.*, 2006, **102**, 5144.
46. M. Pandey, M. C. Iqbal, M Amin, N Ahmad and M. M. Abeer, *Int. J. Polym. Sci.*, 2013, DOI: 10.1155/2013/905471.
47. M. Bhatia and M. Ahuja, *Pharm. Lett.*, 2014, **6**, 127.
48. K. Prasad, G. Mehta, R. Meena and A. K. Siddhanta, *J. Appl. Polym. Sci.*, 2006, **102**, 3654.
49. W. Tanan and S. Saengsuwan, *Energy Procedia*, 2014, **56**, 386.
50. S. K. Bajpai, N. Chand and A. Agrawal, *J. Bioact. Compat. Polym.*, 2015, **30**, 241.
51. L. A. Goetz1, J. R. Sladky and A. J. Ragauskas, *J. Appl. Polym. Sci.*, 2011, **119**, 387.

52. E. L. Rebecca, E. M. Lutton, A. J. Brady, E. M. Vicente-Perez, A. D. Woolfson, R. R. S. Thakur and R. F. Donnelly, *Macromol. Mater. Eng.*, 2015, **300**, 586.
53. E. Khodaverdi, F. Kheirandish, F. S. M. Tekie, B. Z. Khashyarmanesh, F. Hadizadeh and H. M. Haghighi, *ISRN Pharmaceutics*, 2013, Article ID 983053, DOI: 10.1155/2013/983053.
54. S. Pan, X. Yin, Y. F. He, Y. Xiong and R. M. Wang, *Arab. J. Sci. Eng.*, 2015, DOI: 10.1007/s13369-015-1676-z.
55. P. Schexnailder and G. Schmidt, *Colloid Polym. Sci.*, 2009, **287**, 1.
56. N. S. Satarkar and J. Z Hilt, *Acta Biomater.*, 2008, **4**, 11.
57. N. S. Satarkar and J. Z Hilt, *J. Controlled Release*, 2008, **130**, 246.
58. S. Behrens, *Nanoscale*, 2011, **3**, 877.
59. L. Hsu, C. Weder and S. J Rowan, *J. Mater. Chem.*, 2011, **21**, 2812.
60. P. Roy and A. Shahiwala, *J. Controlled Release*, 2009, **134**, 74.
61. http://repository.tamu.edu/handle/1969.1/ETD-TAMU-2010-05-7813.
62. A. P. Alivisatos, *Science*, 1996, **271**, 933.
63. B. Ballou, B. C. Lagerholm, L. A. Ernst, M. P. Bruchez and A. S. Waggoner, *Bioconjugate Chem.*, 2004, **15**, 79.
64. M. Bruchez, P. Moronne, S. Gin Weiss and A. P Alivisatos, *Science*, 1998, **281**, 2013.
65. M. C. Schlamp, X. Peng and A. P. Alivisatos, *J. Appl. Phys.*, 1997, **82**, 5837.
66. S. Chaudhary, M. Ozkan and W. C. W Chan, *Appl. Phys. Lett.*, 2004, **84**, 2925.
67. W. U. Huynh, X. G. Peng and A. P. Alivisatos, *Adv. Mater*, 1999, **11**, 923.
68. A. Y. Nazzal, L. Qu, X. Peng and M. Xiao, *Nano Lett.*, 2003, **3**, 819.
69. G. De Bastida, F. J. Arregui, J. Goicoechea and I. R. Matias, *IEEE Sens. J.*, 2006, **6**, 1378.
70. H. G. Schild, *Prog. Polym. Sci.*, 1992, **17**, 163.
71. K. H. Liu, T. Y. Liu, S. Y. Chen and D. M. Liu, *Acta Biomater.*, 2008, **4**, 1038.
72. S. Menon, M. V. Deepthi, R. R. N. Sailaja and G. S. Ananthapadmanabha, *Indian J. Adv. Chem. Sci.*, 2014, **2**, 76.
73. E. B. Denkbas, *J. Bioact. Compat. Polym.*, 2006, **21**, 351.
74. J. Zhang, H. Chen, L. Xu and Y. Gu, *J. Controlled Release*, 2008, **131**, 34.
75. D. Gao, H. Xu, M. A. Philbert and R. Kopelman, *Nano Lett.*, 2008, **8**, 3320.
76. K. Devi, S Kumar and K. Arivalagan, *Int. J. Pharm. Pharm. Sci.*, 2014, **6**, 118.
77. B. Tyliszczak and K. Pielichowski, *J. Polym. Res.*, 2013, **20**, 191.
78. K. Varaprasad, G. S. M. Redday, J. Jayaramudu, R. Sadiku, K. Ramam, and R. S. Sinha, *Emerging Research Areas and 2013 International Conference on Microelectronics*, Communications and Renewable Energy (AICERA/ICMiCR), 2013 IEEE, pp. 1–4.
79. S. M. Jung, H. Y. Jung, W. Fang, M. S. Dresselhaus and J. Kong, *Nano Lett.*, 2014, **14**, 1810.

Conducting Polymers

7.1 Introduction

The year 1977 saw the discovery of highly conductive polyacetylene by Shirakawa, MacDiarmid and Heeger.[1] This discovery drew attention to π-conjugated polymers as futuristic conducting or semiconducting materials for the next generation of electronic and optical devices. The excellent processability of π-conjugated polymers leads to their deposition by simple printing techniques. The past several decades have witnessed the wide and rapid development of research into conducting polymers, and the initial curiosity about this field has been elevated from academic research to industrial use.[2] These polymers are used in a plethora of applications, such as field-effect transistors (FETs),[3] polymer light-emitting diodes (PLEDs),[4] solar cells[5] and chemical sensors.[6]

The application of microwave (MW) radiation is gaining huge popularity in organic synthesis,[7] especially post the development of single-mode MW sources that allow greater control of both temperature and pressure. As explored in the previous chapters, MW-assisted polymerization is a well-established synthetic tool and is steadily making its mark in the field of conducting polymers as well. Modern MW technology has successfully eliminated problems such as explosions, "runaway" reactions and ineffective heating associated with the traditional multi-mode MW sources. The advent of customized laboratory MW ovens has definitely allowed much better reproducibility of reaction. The most valuable feature of MW irradiated reactions is that it is generally 'quicker and cleaner' than conventional heating, reducing reaction times by more than a factor of 10 and yielding products in high yield with less side-products, high molecular mass and low polydispersity indices.

RSC Green Chemistry No. 35
Microwave-Assisted Polymerization
By Anuradha Mishra, Tanvi Vats and James H. Clark
© Anuradha Mishra, Tanvi Vats and James H. Clark 2016
Published by the Royal Society of Chemistry, www.rsc.org

In this chapter we will discuss the role of MW assistance in the synthesis of conducting polymers with an emphasis on polythiophenes, polyanilne-carbon nanotubes composites (PANI-CNT) and conducting copolymers.

7.2 Synthesis of Polythiophenes

Owing to several useful properties like good charge mobility,[8] efficient fluorescence,[9] high chemical stability and ease of functionality,[10] thiophene oligomers have become important multifunctional materials (Figure 7.1).[11,12] Thiophene oligomers have potential application in thin film transistors,[13] electroluminescent diodes,[14] lasers[15] and photovoltaic cells[9] and as fluorescent markers for biological molecules.[16]

Unfortunately, only very few of these compounds are commercial, the synthesis of the most applications in electronics or biomedical diagnostics are tedious and time-consuming. Initially, polythiophene (PT), was prepared by unsubstituted polythiophene through chemical polymerizations.[17,18] The 2,5-coupled polythiophene that was synthesized was found to be highly conductive and environmentally and thermally stable. The major limitation it had was its insoluble nature. In order to overcome this problem, soluble PTs with flexible side chains, poly(3-alkylthiophene)s (P3ATs) were synthesized by a similar method used for 2,5-coupled polythiophene.[19,20] It is pertinent to note that chemical and electrochemical methods create random couplings in P3ATs, leading to only 50–80% head-to-tail couplings. The loss of regioregularity is due to multiple head-to-head and tail-to-tail couplings, resulting in a sterically twisted structure in the polymer backbone, which leads to a loss of π-conjugation. Steric twisting of backbones leads to destruction of high conductivity and other desirable properties for PTs.

The two most general procedures used to assemble functionalized thienyl rings for the regiospecific synthesis of thiophene oligomers are the Stille[21] and the Suzuki[22] reactions. These reactions are based on the coupling of thienyl metalated reagents with thienyl halogenides or triflates in the presence of palladium or nickel catalysts (Scheme 7.1). These reactions are critically dependent on functionalization types, steric factors, solvent, temperature and the catalyst. The reaction mechanisms are not easy to predict and have to be investigated case by case. This creates great research potential

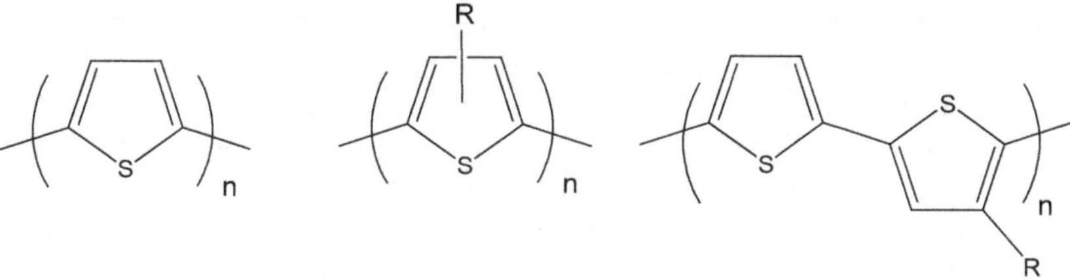

Figure 7.1 Chemical structures of unsubstituted 2,5-coupled poly thiophenes, regioirregular poly(3-alkylthiophene)s (P3ATs), regioregular head-to-tail coupled P3AT.

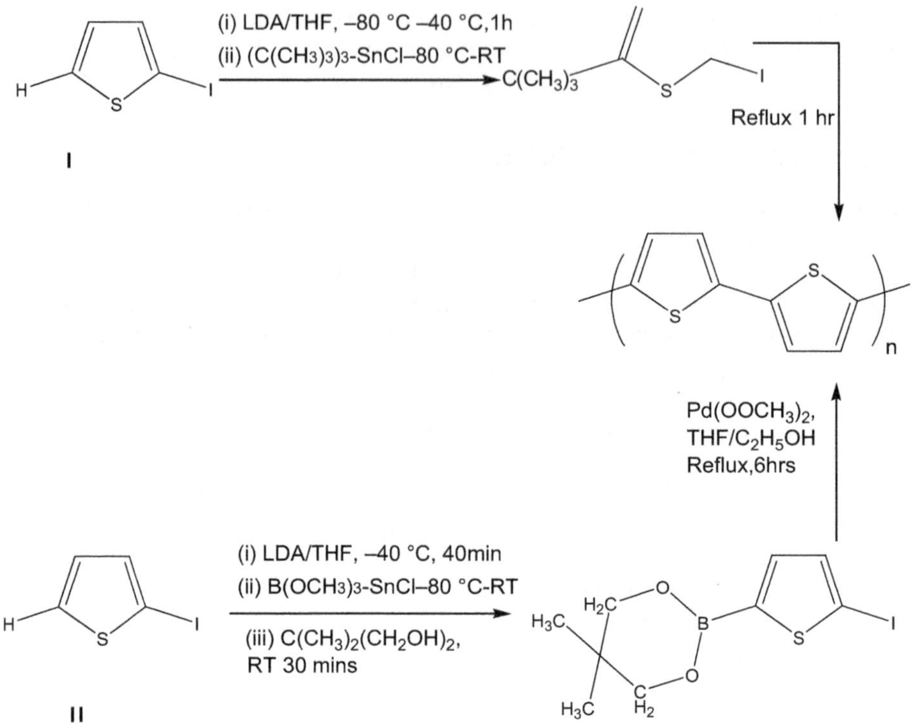

Scheme 7.1 Synthesis of polythiophenes by Stille (ɪ) and Suzuki (ɪɪ) coupling reactions.

for more expedient and efficient synthetic methodologies for the preparation of these oligomers. The optimized reaction conditions ensure the best experimental conditions with minimum byproduct formation and simple purification procedures (in particular, by reducing the huge amounts of organic solvents required to purify the oligomers by silica gel chromatography).[23]

In 2002, Barbarella and co-workers proposed a solvent-free, MW-assisted synthesis of thiophene oligomers *via* the Suzuki coupling reaction.[12] The group based their reaction on the report which suggested that the use of an alumina/potassium fluoride mixture without solvent in the presence of MW irradiation is very effective in palladium catalyzed reactions. They emphasized the effectiveness of the aforementioned procedure in the Suzuki coupling of phenyl iodides with phenylboronic acids.[23] The most important aspect of this reaction is the fact that the absence of solvents makes it possible to recover the reaction products by simple filtration, while the catalyst and the salts formed in the reaction remain on the solid support. On eluting with slightly polar solvents, metal free products are obtained, dramatically reducing the solvent requirement.[24] The authors applied the same procedure for the synthesis of thiophene oligomers. An additional advantage of this synthetic procedure of semiconductors thiophene oligomers is "unintentional doping", *i.e.*, the presence of metallic ions introduced by way of chemical synthesis that alter the intrinsic charge transport properties of the material.[25]

The MW-assisted reactions were carried out in air in a 20 mL reactor using a commercial system Synthewave 402 manufactured by Prolabo. The first and the foremost step involved in the reaction was the catalyst and base optimization. Alumina has always been known for favoring the reaction of adsorbed organic molecules.[26] In order to obtain a reasonably good dispersion of reagents (2,5-dibromothiophene and 2-thiophene boronic acid) and catalyst on the solid support, a mixture of Al_2O_3/reagents/catalyst and a few drops of methanol were mixed (which was subsequently evaporated under reduced pressure). The reaction is depicted in Scheme 7.2. A few attempts were also made with Fluorisil instead of alumina as the solid support, but were unsuccessful. It was also observed that, in some cases, the reaction rate was accelerated by addition of a few drops of aqueous KOH. The experimental conditions were first optimized for the preparation of R-conjugated, unsubstituted terthiophene and then checked in the preparation of unsubstituted quater- and quinque- thiophene, of a dimethylated sexithiophene, of new chiral bithiophenes, and of oligomers containing the thienyl-*S,S*-dioxide moiety.

Different reaction patterns were employed for the preparation of R-conjugated quinque- and sexithiophene. These patterns are summarized in Scheme 7.3.

The authors reported that the careful optimization of the reaction conditions leads to high yields and mixtures that are easy to separate into the different components in just few minutes. Isolated yields of 60–80% can be obtained by choosing the appropriate boron and halogen derivatives. These yields are highly reproducible and competitive, with the best yields already reported for quaterthiophene and quinquethiophene using conventional heating.[26]

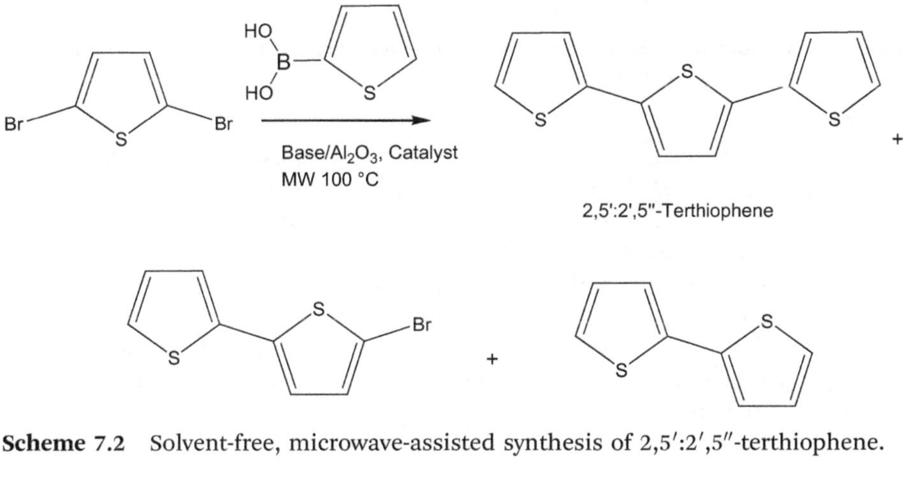

Scheme 7.2 Solvent-free, microwave-assisted synthesis of 2,5':2',5"-terthiophene.

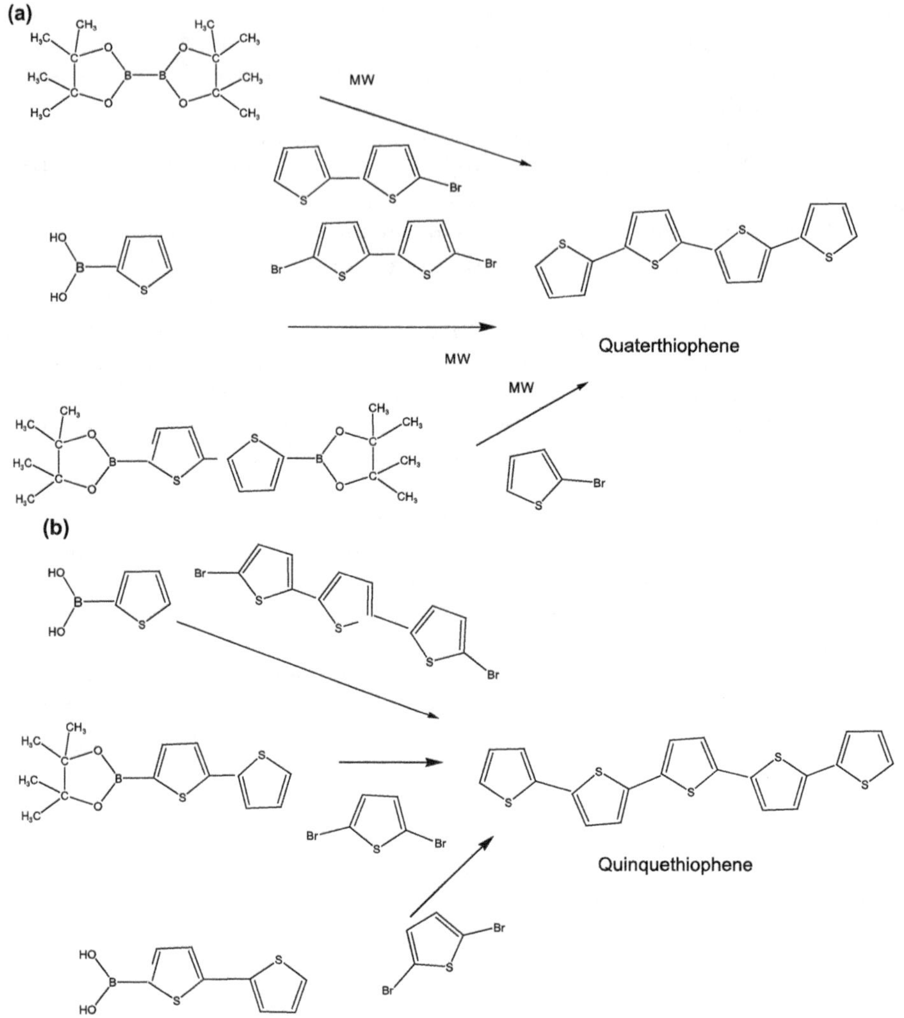

Scheme 7.3 Solvent-free, microwave-assisted synthesis of quaterthiophene and quinquethiophene.

The next step comprised a comprehensive synthesis of 4″,3‴-dimethyl-2,2′ : 5′,2″ : 5′,2‴ :5‴,2″″ : 5″″,2″″″-sexithiophene. The oligomer is a soluble thienyl hexame and in the solid state is nearly planar. It self-assembles in parallel layers, and displays good charge transport properties.[27] In a detailed study, the authors reported a low yield synthesis by means of the Stille coupling.[27] Subsequent higher yields were obtained by means of the Suzuki coupling with the aid of MW, however, it was not a solvent-free reaction.[21] It was after a series of similar reactions the authors reported a MW-assisted solvent free reaction.

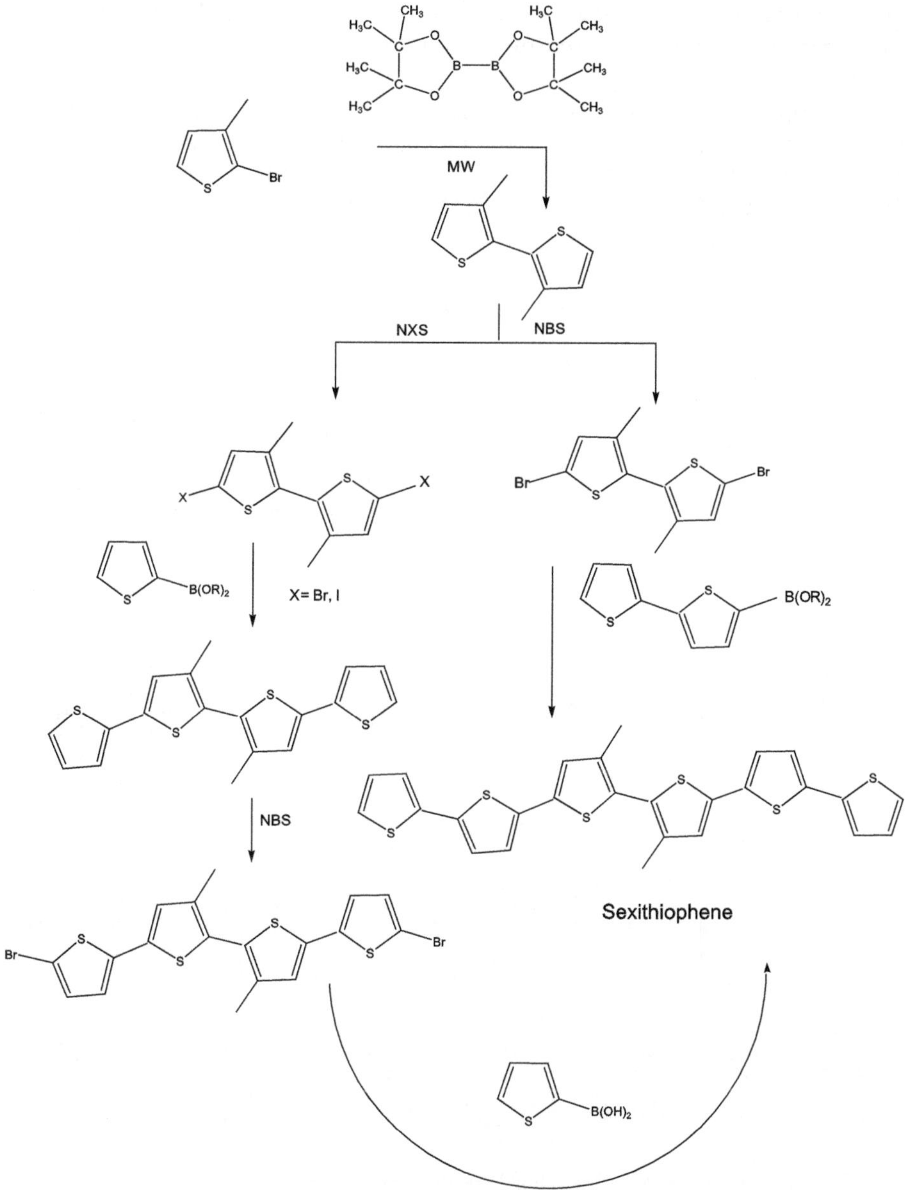

Scheme 7.4 Solvent-free, microwave-assisted synthesis of sexithiophene.

The authors reported that high yields are obtained when one ring is added at a time and the di-iodo derivative is employed, instead of the corresponding di-bromo derivative (Scheme 7.4). This reaction is reported to be highly reproducible and can be viewed as an expedient way to prepare in one pot regioregular head-to-head methyl-substituted sexi- and octathiophenes, which can be readily separated by chromatography.

The same work also gave a detailed MW-assisted synthesis of chiral 2,2′-Bithiophenes in order to investigate the influence of chirality on the self-assembly properties of thiophene oligomers. The synthetic pattern is shown in Scheme 7.5, which shows the monobrominated monomers were obtained by condensation of commercial 5-bromo-2-thiophene aldehyde

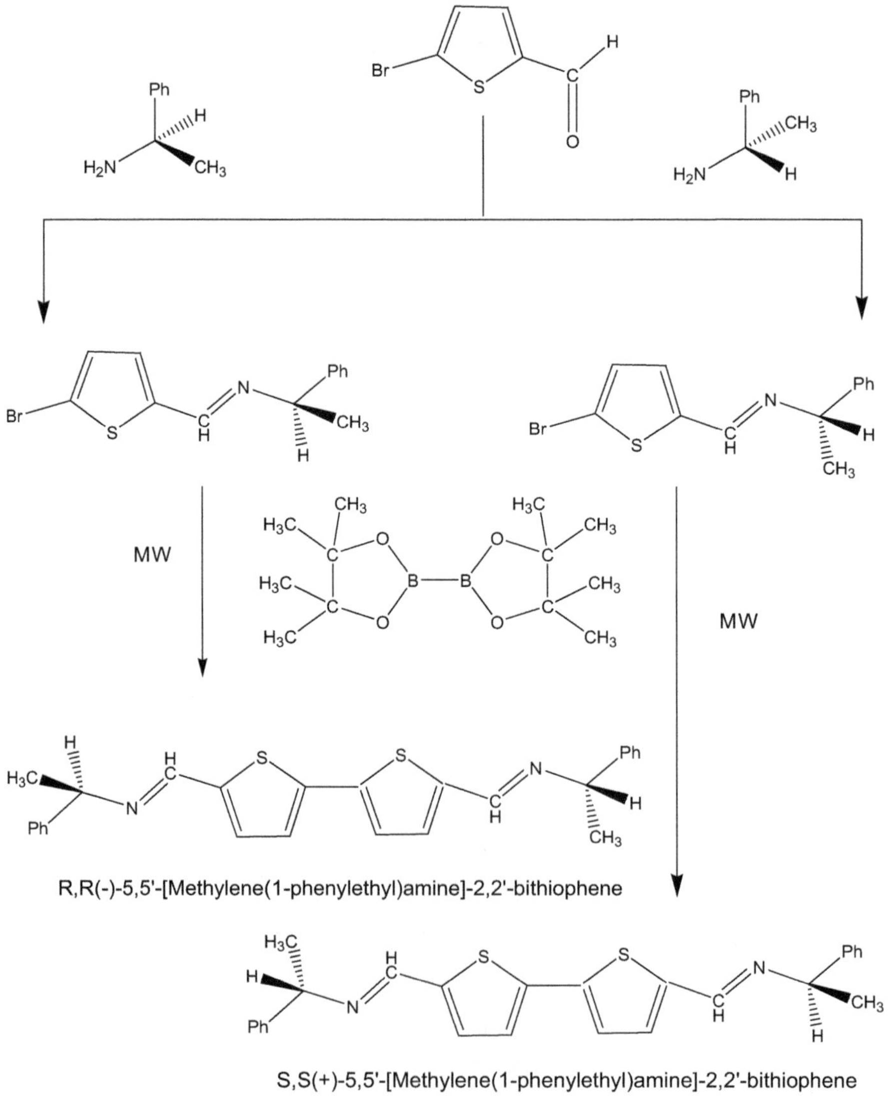

R,R(-)-5,5'-[Methylene(1-phenylethyl)amine]-2,2'-bithiophene

S,S(+)-5,5'-[Methylene(1-phenylethyl)amine]-2,2'-bithiophene

Scheme 7.5 Solvent-free, microwave-assisted synthesis of $R,R(-)$- and $S,S(+)$-5,5′-[Methylene(1-phenylethyl)amine]-2,2′-bithiophenes.

with *R*(−) and *S*(+) 1-phenylethylamine. These monomers further reacted with bis(pinacolato)diboron using the same experimental conditions employed for the preparation of quaterthiophene. A few minutes of MW irradiation resulted in >70% yields of bithiophenes.

The NMR spectrum of the crude product showed the presence of a second iminic peak at 8.43 ppm, indicating that under MW irradiation partial isomerization of the iminic bond (<10%, from 1H NMR) takes place. To exclude the possibility of MW-induced racemization, the optical rotatory dispersion value, α, of monomer *R*(−)-(5-bromothiophen-2-ylmethylene)(1-phenylethyl)amine, was measured before and after MW irradiation. Within the limits of experimental error, the two values ([α]D-133.30° and [α]D-127.38°, respectively) were found to be the same.

The authors also carried out several experiments to get some insight into the reaction mechanisms. The experiments were carried out using thiopheneboronic acid as the starting material, to check the self-coupling reaction of this compound on changing the reaction conditions. The yields estimated by GC/MS analysis are summarized in Figure 7.2. The results show that while no homocoupling is observed in the absence of catalyst even in the presence of MW, the use of microwaves always leads to the formation of dimer, even in the absence of the base.

Tierney and co-workers reported MW-assisted synthesis of polythiophenes *via* the Stille coupling reaction.[28] The authors detailed out the use of MW irradiation in the preparation of soluble polythiophenes *via* the Stille coupling

HO
B
HO
S
KF/Al₂O₃
PdCl₂
S
S
24% without irradiation,
48% after 3min. of MW irradiation
HO
B
HO
S
Al₂O₃
PdCl₂
S
S
No reaction without irradiation,
22% after 3min. of MW irradiation

HO
B
HO
S
KF/Al₂O₃
MW
No reaction

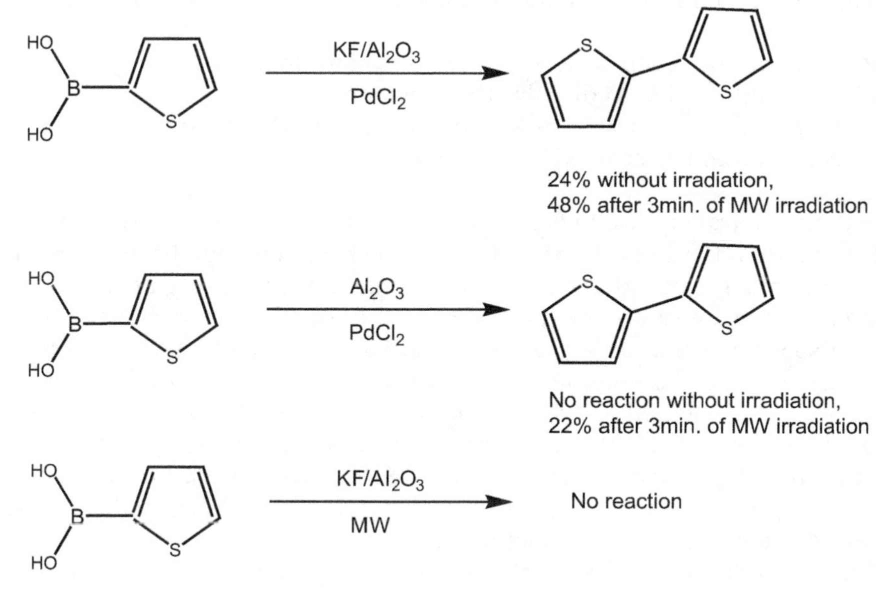

Figure 7.2 Self-coupling reaction of thiophene boronic acid in different experimental conditions.

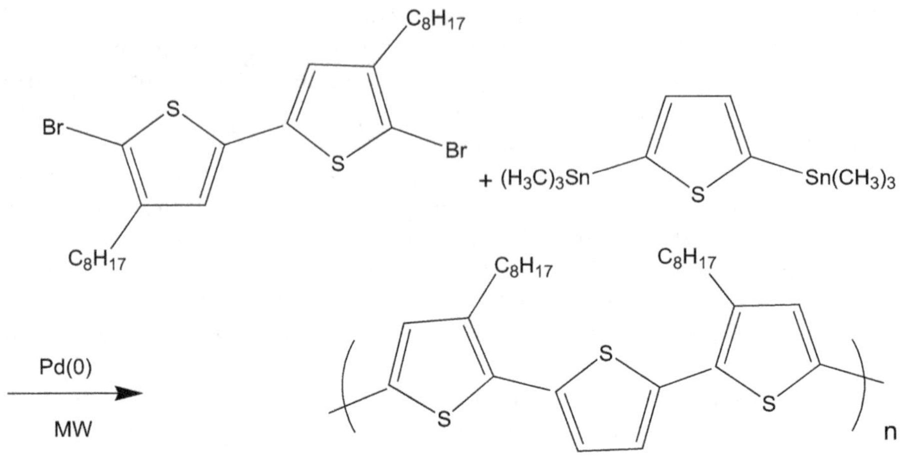

Scheme 7.6 Microwave-assisted synthesis of poly(3,3″-dioctyl-2,2′:5′2″ terthiophene).

with reduced reaction times, increased molecular weights ($Mn > 15\,000$), and low polydispersities ($Mw/Mn \approx 2$). The Stille coupling investigated was between 5,5′-dibromo-4,4′-dioctylbithiophene and 2,5-bis(trimethylstannyl)thiophene to yield poly(3,3′-dioctyl- 2,2′:5′2″-terthiophene), as shown in Scheme 7.6.

The optimization of reaction conditions was done mainly on two parameters, namely, the choice of solvent and the choice of catalyst.

The ideal reaction solvent should possess the following features.

- ✓ It should be compatible with the Stille coupling.
- ✓ It should be able to dissolve the polymer product.
- ✓ It should be amenable to MW heating *i.e.* moderate to high dielectric constant and relatively high boiling.

Although tetrahydrofuran (THF) is able to fully satisfy the first two criteria, it is not a high boiling solvent and cannot be heated using MW radiation to temperatures greater than 140–150 °C, unless there is significant ionic content in the solution. The authors zeroed in on chlorobenzene as the reaction solvent. It met all three criteria, allowing elevated heating temperatures of 200 °C to be obtained using MW radiation. When used as solvent chlorobenzene yielded poly(3,3″-dioctyl-2,2′:5′2″ terthiophene) with improved molecular weights compared to that prepared using THF after a 10 min heating period (Mn 4700 *versus* 1800, respectively). Besides, chlorobenzene proved to be a superior solvent to THF in its ability to dissolve all product material at room temperature.

The catalysts Pd(0) and Pd(ii) both can be reportedly employed in Stille couplings. Although Pd(ii) catalysts are air-stable, they consume organotin reagent when reduced to Pd(0), and so the stoichiometry of the organotin reagent must be accordingly adjusted.[29] Therefore, the authors used only Pd(0) catalysts in order to avoid potential stoichiometry imbalances.[28]

The group also explored the effect of ionic additives, such as Cu(I) salts or LiCl, on the efficacy of Stille couplings.[30] The findings revealed that addition of CuI did not improve the molecular weights of polymer, the addition of LiCl was observed to improve the molecular weight slightly. Varying the ligand/palladium ratio revealed that a higher ligand/palladium ratio did not have a positive effect on the molecular weights. An experiment sans palladium catalyst yielded no polymer. No sign of degradation of either monomer could be found by GC/MS or HPLC analysis even after heating at 200 °C for 20 min.

Iain McCulloch and co-workers studied the influence of molecular design on the field-effect transistor characteristics of terthiophene polymers.[31] In their report, three terthiophene polymers were studied (Figure 7.3).

The authors drew a comparative account of MW-assisted Stille as well as Suzuki coupling and demonstrated that, for the all thiophene systems, attempted Suzuki reactions of thiophene bis(boronic) esters under similar conditions afforded low molecular weight oligomers. This phenomenon was attributed to the possible problems with deboronation.[32] Therefore, the Stille coupling was preferred for the study. The reactions were performed under MW heating to obtain polymers in very short reaction times with typically higher molecular weights than conventional heating.[33,34]

The study concluded that the HOMO energy level, and hence the ionization potential of terthiophene polymers was effectively manipulated through molecular modifications of the polymer backbone, which resulted in a perturbation of the π electron conjugation. These perturbations affect

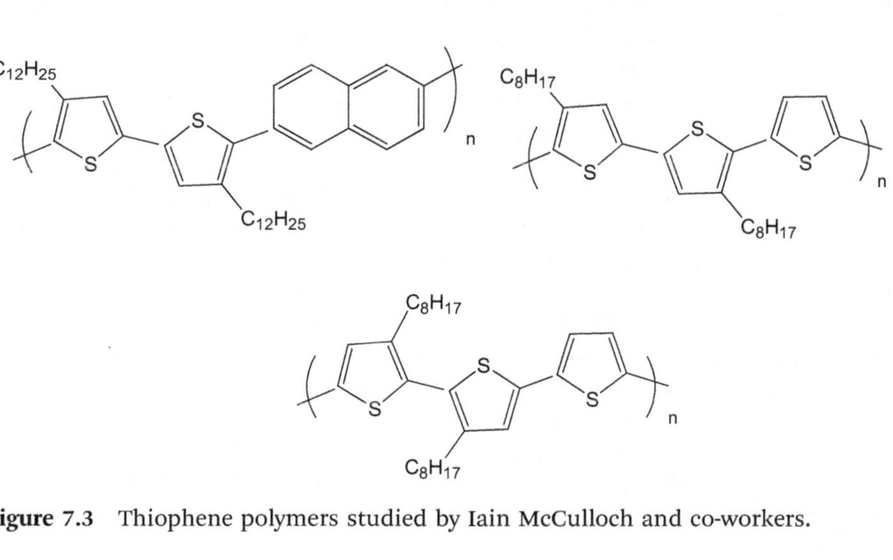

Figure 7.3 Thiophene polymers studied by Iain McCulloch and co-workers.

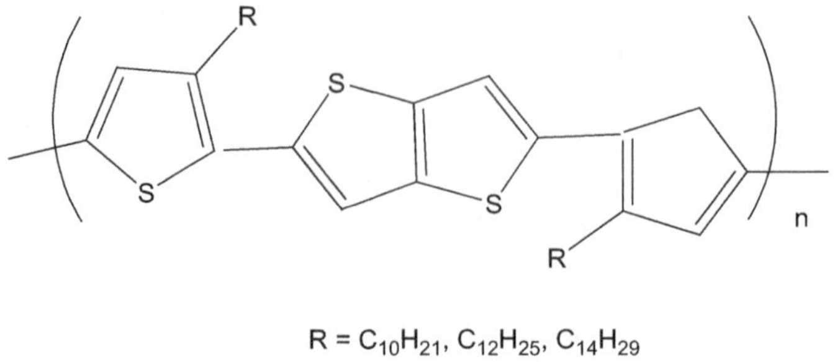

$R = C_{10}H_{21}, C_{12}H_{25}, C_{14}H_{29}$

Figure 7.4   Chemical structure of poly(2,5-bis(3-alkylthiophen-2-yl) thieno[3,2-*b*]thiophene).

both charge carrier mobility and stability of the polymers. A raise in the polymer ionization potential was observed on steric twisting of the thiophene backbone. The twisting, however, led to the loss of close intermolecular aggregation of the thiophene backbone, which in turn had a severely detrimental effect on charge-carrier mobility. Incorporation of the naphthalene group reduced the efficiency of π conjugation along the backbone and increased polymer ionization potential. In this case, some degree of intermolecular π–π aggregation remained, and the reduction in charge carrier mobility was not significant.[29]

The same group reported the MW-assisted synthesis of semiconducting liquid-crystalline thieno[3,2-*b*]thiophene polymers[35] (Figure 7.4).

These liquid crystalline thiophene-based polymers exhibited enhanced charge-carrier mobility, achieved through highly organized morphology from processing in the mesophase. The materials synthesized were characterized using X-ray diffraction and Atomic Force Microscopy. Relatively large crystalline domain sizes on the length scale of lithographically accessible channel lengths (~ 200 nm) were exhibited in thin films. This offered a great potential for fabrication of single-crystal polymer transistors. Good transistor stability under static storage and operation in a low-humidity air environment was demonstrated, with charge-carrier field-effect mobilities of 0.2–0.6 cm^2 V^{-1} s^{-1} achieved under nitrogen. These printable liquid crystalline thiophene-based semiconducting polymers exhibited mobilities. The materials so obtained demonstrated unusual morphological properties relative to other rigid-rod polymers. Besides the relatively large size of the crystalline domains, these materials are well within reach of the nanoscale features that can be fabricated using many lithographic and printing techniques.[36,37] These findings suggest that these materials are indeed an answer to important questions about electrical transport within single-crystalline domains in semiconducting polymers.

C. J. Hawker and co-workers reported the synthesis and characterization of a new family of soluble oligothiophene-*S,S*-dioxides (Figure 7.5) and their application as building blocks to form polythiophene-*S,S*-dioxides *via* MW-assisted Stille coupling polymerization.[38]

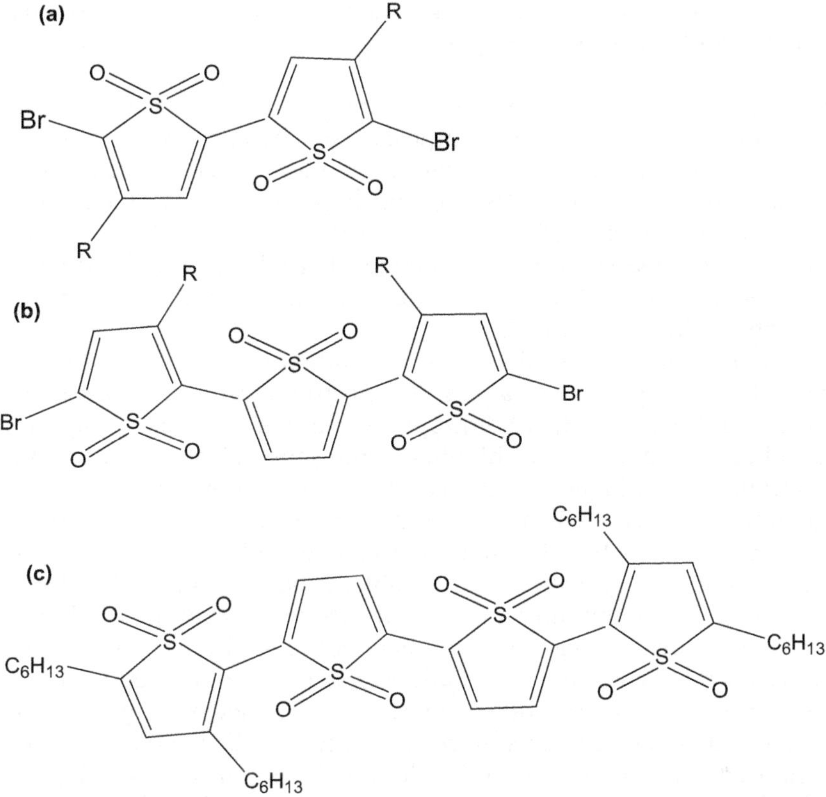

Figure 7.5 Soluble oligothiophene-*S*,*S*-dioxides (a) bithiophene-[all]-*S*,*S* dioxides, (b) terthiophene-[all]-*S*,*S*-dioxides (c) quaterthiophene-[all]-*S*,*S*-dioxide.

MW-assisted reactions were conducted on a Biotage MW reactor at a frequency of 2.5 GHz. Incorporation of hexyl and dodecyl groups into bi-, *tert*-, and quaterthiophene backbones leads to enhanced solubility. The oligothiophene-[all]-*S*,*S*-dioxides also display a reduced energy gap with respect to the aromatic oligothiophene precursors. The solubility of these materials is a boon to their processability and it makes them useful for various applications in organic-based electronic devices. The high solubility also allows them to be used as building blocks for the preparation of thiophene-based conjugated polymers containing bithiophene-*S*,*S*-dioxide units in the backbone. Incorporation of electron-withdrawing sulfones along the polymer backbone leads to the narrowing of the energy gap with the energies of both frontier orbitals in bithiophene-[all]-*S*,*S*-dioxide containing polymers being lower with respect to the aromatic polythiophene. The resultant materials have a great electron-accepting ability tendency as the narrowing of the energy gap is considerably stronger for the LUMO than for the HOMO. The modularity of this approach potentially allows the energy levels of polythiophenes to be controlled, based on the number and position of the thiophene-*S,S*-dioxide units, thus making them useful for small molecule- and polymer-based devices, such as thin film transistors.

In a quest to obtain fine-tuning of homo levels, Kleinhenz and co-workers experimented with MW-assisted synthesis of low-band-gap thienothiophene-based polymers that utilize quinoid resonance structure stabilization.[39] With the aid of the CEM Discover Benchmate microwave reactor they synthesized new series of polymers based on a prequinoid structure, thieno[3,4-b]thiophene, and a comonomer, benzo[1,2-b:4,5-b']dithiophene. These moieties attained over 7% power conversion efficiency in bulk heterojunction polymer solar cells.

In order to explore the utility of the thienothiophene, the authors studied a library of six polymers by varying the electronic properties of the comonomers (naphtha[2,1-b:3,4-b']dithiophene, dithieno-[3,2-f:2',3'-h]quinoxaline, and benzo[1,2-b:4,5-b']dithiophene) and those of the thienothiophene. The group, interestingly, discovered that the thienothiophene unit predominantly decides the low-band-gap characteristic of these polymers. It was also comprehended that the highest occupied molecular orbital (HOMO) energy level of these polymers was tunable, depending upon the electronic properties of the comonomer and the substitution of fluorine on thiophene moiety. Therefore, the open-circuit voltage of related bulk hetero junction devices changes accordingly. The authors also observed that all short circuit currents were low, which limited the overall efficiency of all devices to less than 1%. Such low currents were attributed to low molecular weight, unoptimized side chains, and low hole mobilities. All these findings point out the fact that material optimization to achieve high efficiency polymer solar cells is a convoluted process; side chain length (and shape) and molecular weight, in addition to band gap and energy levels, all play a crucial and interdependent role.

7.3 Synthesis of Polyaniline-carbon Nanotubes (PANI-CNT) Composites

Because of their charge, storage capacity super capacitors have a great potential application in electrical vehicles, portable computers and cellular devices.[40–43] Carbon nanotubes (CNTs) are frequently studied for super capacitor applications because of their unique electric properties and nanoscale size.[44–46] But CNTs possess lower capacitance than many electronically conducting polymers.[47,48] On the other hand, many studies have proved that electronically conducting polymers represent promising pseudo-capacitive materials due to versatility of structure, low cost and high specific capacitance.[49–52] Thus, the combination of electronically conducting polymers with CNTs offers an attractive way to improve the capacitance of materials resulting from the large specific surface of CNTs and redox contribution of the conducting polymers.[53–57]

Chemical or electrochemical polymerization is a standard efficient tool preparing the conducting polymers/CNTs composite. Zhang and co-workers reported MW-assisted synthesis of the polyaniline/multi-wall carbon nanotubes (PANI/MWNTs) composite and studied their capacitance.[58] The advantages of this method include quick thermal reactions, low energy consumption and the controllable thickness of polymer layers.[59–61]

Using a commercial microwave oven equipped with a reflux condenser Zhang *et al.* synthesized PANI/MWNTs composite. Purified MWNTs and aniline were dispersed in 0.2 M H_2SO_4 was sonticated, to which ammonium persulfate was added. Then the mixture was heated using a commercial MW oven for three minutes, and then cooled to room temperature. The resulting powders were filtered and rinsed with deionized water and ethanol several times, and then dried at 50 °C overnight under vacuum conditions, to obtain the composites.

The resulting composites were characterized by FTIR and TEM and the electrochemical tests indicated that this composite had a high specific capacitance (322 F/g) and good rate capability.

7.4   Conducting Copolymers

Low-cost organic thin-film transistor circuits for flexible electronics require printable, solution-processable organic semiconductors with relatively high mobility.[62–67] Thiophene-based conjugated polymers have been extensively researched as materials for such applications.[68–71] In particular, as a result of the structural regularity of the polymer backbone, regioregular poly(3-hexylthiophene) exhibits a relatively high charge-carrier (hole) mobility.[72,73]

From the previous sections, it is clear that the morphology plays an important role in the electronic and optical properties of conducting polymers. It is thus pertinent that controlling morphology plays an important role in the device application aspect of conducting polymers. Common methods of controlling morphology comprise blending, block copolymer approaches, conjugate polymers synthesis, solvent and thermal annealing, as well as hard templating approaches.[74–80] Most of these approaches are based on the utilization of donor–acceptor architectures within the backbone. The donor–acceptor systems cause partial intramolecular charge transfer that enables manipulation of the electronic structure (HOMO/LUMO levels), leading to low band gap semiconducting polymers[81–86] with relatively high charge carrier mobilities.[87–94]

In order to create a push–pull system, a series of polymers based on diaryl-dithienylcyclopentadienone (2,5-dithienyl-3,4-(1,8-naphthylene) cyclopentadienone, (DTCPD) and thiophene materials, revealing a broad absorption band covering the whole visible region has been prepared (Figure 7.6).[95–97] 4,4′-dialkyl-[2,2′]bithiophene (DAT) moieties are selected as an electron-rich comonomer for DTCPD since tail-to-tail regiopositioning of the alkyl chains on the thiophene monomer helps promote self-organization, while minimizing any steric interactions between neighboring alkyl groups, thus preserving backbone planarity.[98] Besides, the

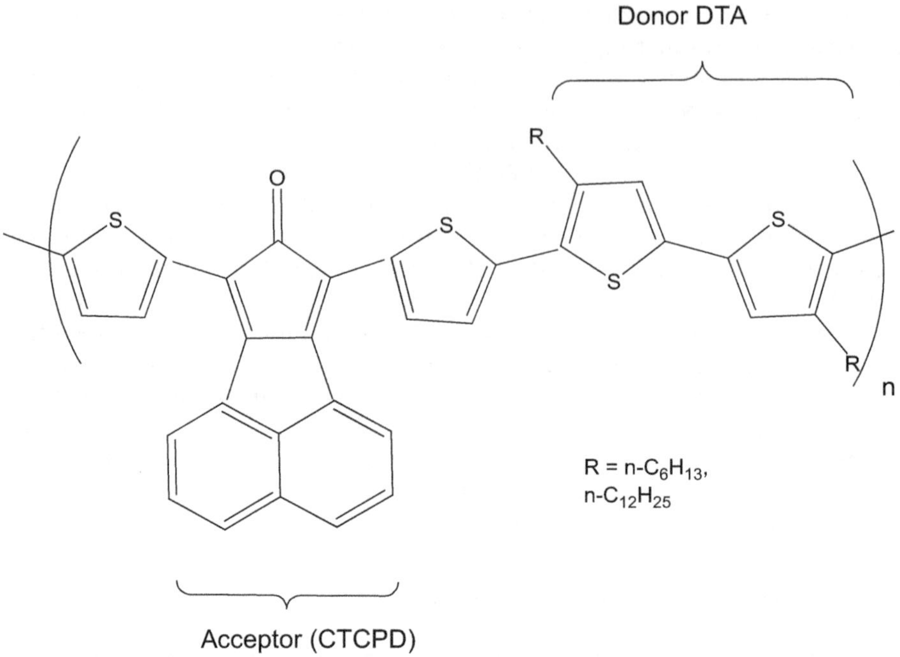

Figure 7.6 Chemical structure of alternate DCTPD and thiophene polymers creating a push–pull electronic system.

unsubstituted thiophene units in DTCPD can increase the ionization potential by improving rotational freedom,[99] resulting in enhanced oxidative stability when compared to that of the poly(alkylthiophene)s.

The authors reported a comparative synthesis of the DTCPD-DAT-based copolymers facilitated by MW and conventional heating. Although both DTCPD-*alt*-DHT and DTCPD*alt*-DDT were successfully synthesized by using a MW-assisted heating protocol, the polymerization fails to yield high molecular weight copolymers (Mn) 4500–6000 g mol^{-1}). The reason for the low molecular weights is likely the poor solubility of the thiophene-based monomer and/or reduced stability of the intermediate palladium complexes involving DTCPD units (Scheme 7.7).

Weber and Thomas reported the synthesis of Poly(*p*-phenylene) and Poly(phenyleneethynylene) conjugate polymers based on a spirobifluorene building block.[100] The synthetic steps involved 2,2′,7,7′-Tetrabromo-9,9′-spirobifluorene as the structure directing monomer in the synthesis of three different conjugated polymers. The tetrahedral carbon atom directs the two fluorene moieties into a cross-like shape, giving rise to a contorted polymer structure.

The synthetic procedure involved reaction of 2,2′,7,7′-Tetrabromo-9,9′-spirobifluorene with benzene-1,4-diboronic acid in a mixture of *N,N*-dimethylformamide and water (20 vol %) under the presence of suitable catalyst (Scheme 7.8). The MW-assisted synthesis successfully led to quantitative and

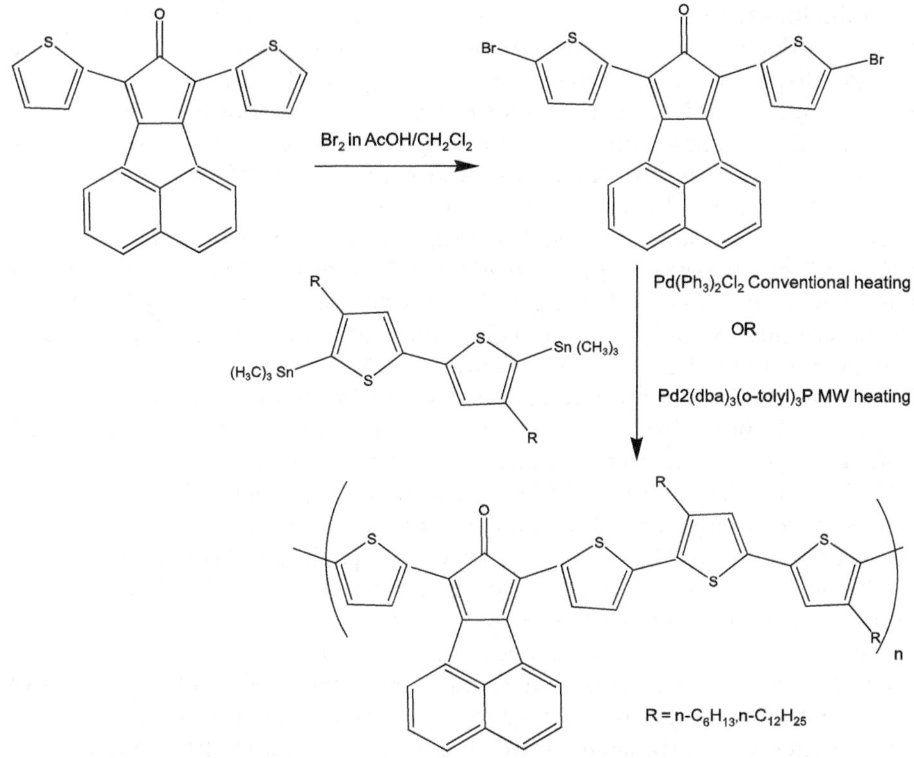

Scheme 7.7 Synthesis of alternate DCTPD and thiophene polymers.

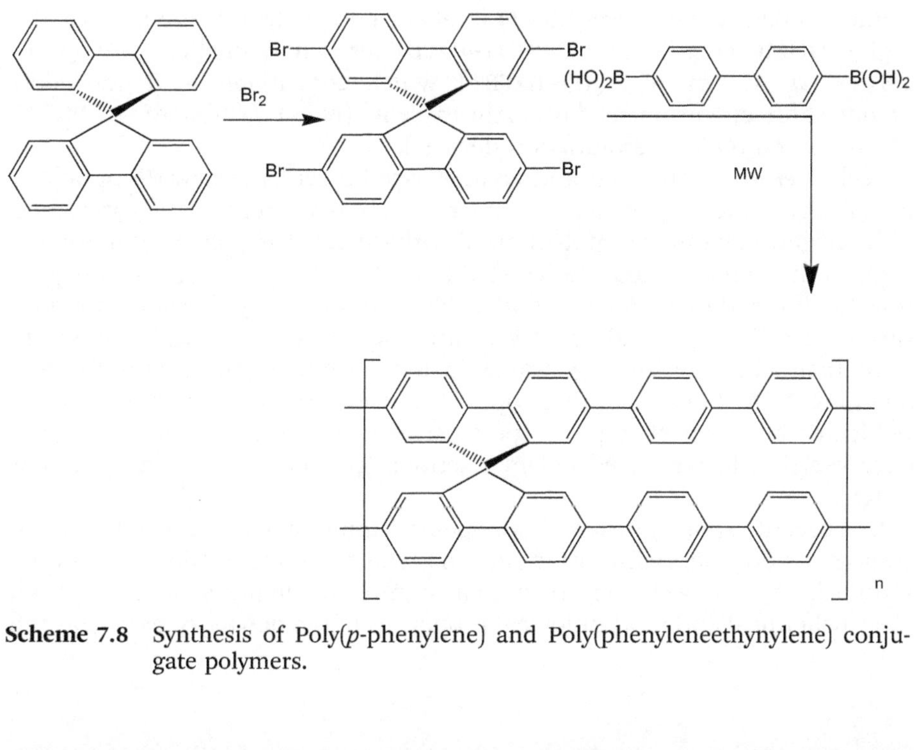

Scheme 7.8 Synthesis of Poly(*p*-phenylene) and Poly(phenyleneethynylene) conjugate polymers.

reproducible synthesis of the *p*-phenylene type networks. Characterization of the networks was accomplished by IR spectroscopy, elemental analysis and energy dispersive X-ray spectroscopy (EDX). With the help of photo-luminescence studies, the authors were able to show that conjugated polymers based on a spirobifluorene building unit possess microporosity. This approach allows the introduction of a large, stable interface into materials with great potential for use in organic electronics.

You and co-workers proposed a design strategy of "weak donor-strong acceptor" copolymer to approach ideal polymers with both a low HOMO energy level and a small band gap.[101] This design is particularly important for heterojunction polymer solar cells as it results in both high open circuit voltage and high short circuit current.

In order to fully comprehend the delicate interplay of the open circuit voltage and short circuit current through molecular design of conjugated polymers, two such "weak donors", naphtha[2,1-b:3,4-b′]- dithiophene (NDT) and dithieno[3,2-f:2′,3′-h]quinoxaline (QDT), which differ only by two atoms, were copolymerized with a common acceptor, 4,7-di(2-thienyl)-2,1,3-benzothiadiazole (DTBT) in a CEM Discover Benchmate MW reactor.

The bulk heterojunction devices based on PNDT-4DTBT and PQDT-4DTBT as the donor polymer (with PC61BM as the acceptor) exhibit overall efficiencies of 5.1% and 4.3%, respectively. With the help of two structurally related polymers, it was established that incorporating electron-withdrawing atoms in the donor unit would lead to a lower HOMO energy level. This lower HOMO energy level translates into a higher open circuit voltage of 0.83 V in PQDT-4DTBT-based bulk heterojunction devices. In contrast, the slightly higher HOMO energy level (-5.34 eV) of PNDT-4DTBT limits the open circuit voltage to 0.67 V. Since the LUMO levels of both polymers are similar, the higher HOMO energy level of PNDT-4DTBT leads to a smaller band gap of 1.61 eV (*vs.* 1.70 eV of PQDT-4DTBT), which contributes to a noticeably higher short circuit current of 14.20 mA/cm^2 (*vs.* 11.38 mA/cm^2 of PQDT-4DTBT based bulk heterojunction devices).

Leclerc *et al.* synthesized and characterized a series of new dithieno[3,2-b:2′,3′-d]germole copolymers.[102] In the extensive report the presented dithienogermole unit was polymerized with different aromatic comonomers such as the benzothiadiazole (PGe1-C8 and PGe1-EH) and thieno[3,4-c]pyr-role-4,6-dione (PGe2). Suzuki and Stille coupling polymerizations under various conditions including MW irradiations were utilized. The polymers were then characterized by size-exclusion chromatography and thermal analyses (TGA, DSC), and their optical and electronic properties were investigated by UV-vis absorption spectroscopy and cyclic voltammetry. It was observed that MW-assisted polymerizations led to an increased molecular weight.

In a recent report a series of cyclopentadithiophene-based low band gap conjugated polymers with varied side-chain patterns and F-substituents were synthesized.[103] It were observed that significant changes in the optical, electrochemical and morphological properties of the polymers, as well as the subsequent performance of devices were attained by replacing the shorter 2-ethylhexyl side-chain with the longer 3,7-dimethyloctyl side-chain. The authors utilized MW-assisted Stille coupling polymerization for the synthesis of the conjugated polymers 2-ethylhexyl side-chain and monofluoro substituent.

Solar cells fabricated from these conjugated polymers exhibit increased open circuit voltage, short circuit current and fill factor, resulting in the highest power conversion efficiency (5.5%) in this series of polymers. This finding provides valuable insight for making more efficient low band gap polymers.

Zhang and co-workers synthesized highly fluorescent polymerizable carbon nanodots by MW-assisted pyrolysis and subsequent surface vinylation.[104] These nanodots were copolymerized with several model monomers to form different functional fluorescent polymeric materials in solution or the solid-state, indicating a simple and versatile approach to novel fluorescent polymer materials.

Bensebaa and co-workers utilized MW heating to prepare platinum-ruthenium nanoparticles stabilized within a Polypyrrole di(2-ethylhexyl) sulfosuccinate matrix.[105] These composites exhibited great electrical conductivity, thermal stability and solubility in polar organic solvents. The method comprised a scalable and quick two-step process to fabricate alloyed nanoparticles dispersed in the matrix. Physical characterization by TEM revealed that crystalline and monodisperse alloyed nanoparticles had an average size of 2.8 nm. With the help of methanol electro-oxidation data, the authors proposed these materials as potential candidates for direct methanol fuel cell application.

Another MW-assisted synthesis of composites is reported for Polypyyrole/Ag nano-comoposites.[106] Interfacial polymerization was employed to synthesize the nanocomposites in a water/chloroform interface. The merit of the process lies in the fact that MW-aided synthesis does not require any oxidizing agent for polymerization. The formation of Polypyyrole was confirmed by absorption spectroscopy showing a band at $\sim$464 nm, whereas XRD measurement was employed to confirm the formation of Polypyyrole/Ag nanocomposite.

A similar example of interfacial polymerization is self-assembled copolymerization of poly(anilineco-*p*-phenylenediamine) nanorods *via* aqueous/ionic liquid interfacial oxidative polymerization in the presence of acid dopants.[107] The method proved to be an easy and economical non-template route for synthesizing copolymer nanostructures. The morphology of copolymer nanostructures so formed was investigated using FE-SEM and FE-TEM, which indicate the formation of one-dimensional (1D) nanorods with an average diameter of 45–100 nm. FTIR, XRD and UV–visible measurements confirmed the molecular and electrical structure of the doped copolymer. An important feature of this reaction is that the electrical conductivity of copolymer nanorods depends upon the nature of dopant used.

We can hence conclude that MW-assisted synthesis of conducting polymers will definitely be a breakthrough in green chemistry. These reactions can rightly be called doubly blessed as they not only ensure solvent-free synthesis, the products generated (conducting polymers) are largely utilized in bulk heterojunction solar cells, leading to green energy.

References

1. C. K. Chiang, C. R. Fincher, Y. W. Park Jr., A. J. Heeger, H. Shirakawa, E. J. Louis, S. C. Gau and A. G. MacDiarmid, *Phys. Rev. Lett.*, 1977, **39**, 1098.
2. I. Osaka and R. D. Mccullough, *Acc. Chem. Res.*, 2008, **41**, 1202.
3. H. Sirringhaus, N. Tessler and R. H. Friend, *Science*, 1998, **280**, 1741.
4. M. R. Andersson, O. Thomas, W. Mammo, M. Svensson, M. Theander and O. S. Ingana, *J. Mater. Chem.*, 1999, **9**, 1933.
5. G. Li, V. Shrotriya, J. Huang, Y. Yao, T. Moriarty, K. Emery and Y. Yang, *Nat. Mater.*, 2005, **4**, 864.
6. D. T. McQuade, A. E. Pullen and T. M. Swager, *Chem. Rev.*, 2000, **100**, 2537.
7. P. Lidstrom, J. Tierney, B. Wathey and J. Westman, *Tetrahedron*, 2001, **57**, 9225.
8. Y. Kanemitsu, K. Suzuki, Y. Masumoto, Y. Tomiuchi, Y. Shiraishi and M. Kuroda, *Phys. Rev. B: Condens. Matter Mater. Phys.*, 1994, **50**, 2301.
9. F. Garnier, *Acc. Chem. Res.*, 1999, **32**, 209.
10. H. Chosrovian, S. Rentsch, D. Grebner, U. Dahom and E. Birckner, *Synth. Met.*, 1993, **60**, 23.
11. P. Blanchard, B. Jousselme, P. Frère and J. Roncali, *J. Org.Chem.*, 2002, **67**, 3961.
12. M. Melucci, G. Barbarella and G. Sotgiu, *J. Org. Chem.*, 2002, **67**, 8877.
13. C. D. Dimitrakopoulos and D. J. Mascaro, *IBM J. Res., Dev.*, 2001, **45**, 11.
14. M. Suzuki, M. Fukuyama, Y. Hori and S. Hotta, *J. Appl. Phys.*, 2002, **91**, 5706.
15. M. Zavelani-Rossi, G. Lanzani, S. De Silvestri, M. Anni, G. Gigli, R. Cingolani, G. Barbarella and L. Favaretto, *Appl. Phys. Lett.*, 2001, **79**, 4082.
16. M. Catellani, B. Boselli, S. Luzzati and C. Tripodi, *Thin Solid Films*, 2002, **403**, 66.
17. J. J. Apperloo, C. Martineau, P. A. van Hal, J. Roncali and R. A. J. Janssen, *J. Phys. Chem. A*, 2002, **106**, 21.
18. G. Barbarella, M. Zambianchi, O. Pudova, V. Paladini, A. Ventola, F. Cipriani, G. Gigli, R. Cingolani and G. Citro, *J. Am. Chem. Soc.*, 2001, **123**, 11600.
19. T. Yamamoto, K. Sanechika and A. Yamamoto, *J. Polym. Sci., Polym. Lett. Ed.*, 1980, **18**, 9.
20. J. W. P. Lin and L. P. Dudek, *J. Polym. Sci., Polym. Chem. Ed.*, 1980, **18**, 2869.

21. G. Li, V. Shrotriya, J. Huang, Y. Yao, T. Moriarty, K. Emery and Y. Yang, *Nat. Mater.*, 2005, **4**, 864.
22. D. T. McQuade, A. E. Pullen and T. M. Swager, *Chem. Rev.*, 2000, **100**, 2537.
23. D. Villemin and F. Caillot, *Tetrahedron Lett.*, 2001, **42**, 639.
24. G. W. Kabalka, R. M. Pagni and C. M. Hair, *Org. Lett.*, 1999, **1**, 1423.
25. R. Hajlaoui, G. Horowitz, F. Garnier, A. Arce-Brouchet, L. Laigre, A. El Kassmi, F. Demanze and F. Kouki, *Adv. Mater.*, 1997, **9**, 389.
26. A. Merz and F. Ellinger, *Synthesis*, 1991, **6**, 462.
27. G. Barbarella, M. Zambianchi, L. Antolini, P. Ostoja, P. Maccagnani, A. Bongini, E. A. Marseglia, E. Tedesco, G. Gigli and R. Cingolani, *J. Am. Chem. Soc.*, 1999, **121**, 8920.
28. S. Tierney, M. Heeney and I. McCulloch, *Synth. Met.*, 2005, **148**, 195.
29. Z. Bao, W. K. Chan and L. Yu, *J. Am. Chem. Soc.*, 1995, **117**, 12426.
30. V. Farina, V. Krishnamurthy and W. J. Scott, *Org. React.*, 1997, **50**, 1.
31. I. McCulloch, C. Bailey, M. Giles, M. Heeney, I. Love, M. Shkunov, D. Sparrowe and S. Tierney, *Chem. Mater.*, 2005, **17**, 1381.
32. M. Jayakannan, L. Joost, J. van Dongen and R. A. J. Janssen, *Macromolecules*, 2001, **34**, 5386.
33. B. S. Nehls, U. Asawapirom, S. Fuldner, E. Preis, T. Farrell and U. Scherf, *Adv. Funct. Mater.*, 2004, **14**, 352.
34. T. Sano, Y. Hamada and K. Shibata, *IEEE J. Sel. Top. Quantum Electron.*, 1998, **4**, 34.
35. I. McCulloch, M. Heeney, C. Bailey, K. Genevicius, I. MacDonald, M. Shkunov, D. Sparrowe, S. Tierney, R. Wagner, W. Zhang, M. L. Chabinyc, R. J. Kline, M. D. McGehee and M. F. Toney, *Nat. Mater.*, 2006, **5**, 328.
36. Y. Xia, J. A. Rogers, K. E. Paul and G. M. Whitesides, *Chem. Rev.*, 1999, **99**, 1823.
37. C. W. Sele, T. vonWerne, R. H. Friend and H. Sirringhaus, *Adv. Mater.*, 2005, **17**, 997.
38. E. Amir, K. Sivanandan, J. E. Cochran, J. J. Cowart, S. Yu, Ku, J. H. Seo, M. L. Chabinyc and C. J. Hawker, *J. Polym. Sci., Part A: Polym. Chem.*, 2011, **49**, 1933.
39. N. Kleinhenz, L. Yang, H. Zhou, S. C. Price and W. You, *Macromolecules*, 2011, **44**, 872.
40. B. E. Conway, V. Birss and J. Wojtowicz, *J. Power Sources*, 1997, **66**, 1.
41. J. N. Broughton and M. J. Brett, *Electrochim. Acta*, 2004, **49**, 4439.
42. V. Gupta, T. Kusahara, H. Toyama, S. Gupta and N. Miura, *Electrochem. Commun.*, 2007, **9**, 2315.
43. R. Kotz and M. Carlen, *Electrochim. Acta*, 2004, **45**, 2483.
44. Y. G. Wang, H. Q. Li and Y. Y. Xia, *Adv. Mater.*, 2006, **18**, 2619.
45. F. Béguin, K. Szostak, G. Lota and E. Frackowiak, *Adv. Mater.*, 2005, **17**, 2380.
46. J. Okuno, K. Maehashi, K. Matsumoto, K. Kerman, Y. Takamura and E. Tamiya, *Electrochem. Commun.*, 2007, **9**, 13.

47. M. Ishikawa, M. Morita, M. Ihara and Y. Matsuda, *J. Electrochem. Soc.*, 1994, **141**, 1730.
48. S. T. Mayer, R. W. Pekala and J. L. Kascchmitter, *J. Electrochem. Soc.*, 1993, **140**, 446.
49. M. Hughes, G. Z. Chen, M. S. P. Shaffer, D. J. Fray and A. H. Windle, *Chem. Mater.*, 2002, **14**, 1610.
50. L. Z. Fan and J. Maier, *Electrochem. Commun.*, 2006, **8**, 937.
51. C. Arbizzani, M. C. Gallazzi, M. Mastragostino, M. Rossi and F. Soavi, *Electrochem. Commun.*, 2001, **3**, 16.
52. G. A. Snook, C. Peng, D. J. Fray and G. Z. Chen, *Electrochem. Commun.*, 2007, **9**, 83.
53. M. Hughes, M. S. P. Shaffer, A. C. Renouf, C. Singh, G. Z. Chen, D. J. Fray and A. H. Windle, *Adv. Mater.*, 2002, **14**, 382.
54. D. Wei, C. Kvarnstrom, T. Lindfors and A. Ivaska, *Electrochem. Commun.*, 2007, **9**, 206.
55. Z. Wei, M. Wan, T. Lin and L. Dai, *Adv. Mater.*, 2003, **15**, 136.
56. B. Philip, J. Xie, J. K. Abraham and V. K. Varadan, *Polym. Bull.*, 2005, **53**, 127.
57. G. M. Milena, G. M. Janis, C. Raoul, P. S. George and M. F. Peter, *Chem. Mater.*, 2006, **18**, 6258.
58. H. Mi, X. Zhang, S. An, X. Ye and S. Yang, *Electrochem. Commun.*, 2007, **9**, 2859.
59. J. Palacios and C. Valverde, *New Polym. Mater.*, 1996, **5**, 93.
60. P. K. D. V. Yarlagadda and T. C. Chai, *J. Mater. Process. Technol.*, 1998, **74**, 199.
61. M. R. Karim, C. J. Lee, A. M. S. Chowdhury, N. Nahar and M. S. Lee, *Mater. Lett.*, 2007, **61**, 1688.
62. H. Chosrovian, S. Rentsch, D. Grebner, U. Dahom and E. Birckner, *Synth. Met.*, 1993, **60**, 23.
63. Z. N. Bao, *Adv. Mater.*, 2000, **12**, 227.
64. S. R. Forrest, *Nature*, 2004, **428**, 911.
65. H. Sirringhaus, *Adv. Mater.*, 2005, **17**, 2411.
66. H. Sirringhaus, T. Kawase, R. H. Friend, T. Shimoda, M. Inbasekaran, W. Wu and E. P. Woo, *Science*, 2000, **290**, 2123.
67. H. Sirringhaus, N. Tessler and R. H. Friend, *Science*, 1998, **280**, 1741.
68. R. D. McCullough, *Adv. Mater.*, 1998, **10**, 93.
69. Z. Bao, A. Dodabalapur and A. J. Lovinger, *Appl. Phys. Lett.*, 1996, **69**, 4108.
70. R. D. McCullough, S. Tristramnagle, S. P. Williams, R. D. Lowe and M. Jayaraman, *J. Am. Chem. Soc.*, 1993, **115**, 4910.
71. D. Fichou and C. Ziegler, in *Handbook of Oligothiophenes and Polythiophenes*, ed. D. Fichou, Wiley-VCH, Weinheim, Germany, 1999, ch. 4, p. 183.
72. H. Sirringhaus, P. J. Brown, R. H. Friend, M. M. Nielsen, K. Bechgaard, B. M. W. Langeveld-Voss, A. J. H. Spiering, R. A. J. Janssen, E. W. Meijer, P. Herwig and D. M. de Leeuw, *Nature*, 1999, **401**, 685.

73. T. J. Prosa, M. J. Winokur, J. Moulton, P. Smith and A. J. Heeger, *Macromolecules*, 1992, **25**, 4364.

74. A. C. Grimsdale and K. Mullen, *Adv. Polym. Sci.*, 2006, **199**, 1.

75. A. P. Kulkarni, C. J. Tonzola, A. Babel and S. A. Jenekhe, *Chem. Mater.*, 2004, **16**, 4556.

76. G. Li, V. Shrotriya, J. S. Huang, Y. Yao, T. Moriarty, K. Emery and Y. Yang, *Nat. Mater.*, 2005, **4**, 864.

77. W. L. Ma, C. Y. Yang, X. Gong, K. Lee and A. J. Heeger, *Adv. Funct. Mater.*, 2005, **15**, 1617.

78. M. M. Alam and S. A. Jenekhe, *Macromol. Rapid Commun.*, 2006, **27**, 2053.

79. J. W. Andreasen, M. Jorgensen and F. C. Krebs, *Macromolecules*, 2007, **40**, 7758.

80. X. Yang and J. Loos, *Macromolecules*, 2007, **40**, 1353.

81. Q. T. Zhang and J. M. Tour, *J. Am. Chem. Soc.*, 1997, **119**, 5065.

82. A. K. Agrawal and S. A. Jenekhe, *Chem. Mater.*, 1996, **8**, 579.

83. T. Yamamoto, Z. H. Zhou, T. Kanbara, M. Shimura, K. Kizu, T. Maruyama, Y. Nakamura, T. Fukuda, B. L. Lee, N. Ooba, S. Tomaru, T. Kurihara, T. Kaino, K. Kubota and S. Sasaki, *J. Am. Chem. Soc.*, 1996, **118**, 10389.

84. J. P. Ferraris, A. Bravo, W. Kim and D. C. Hrncir, *J. Chem. Soc., Chem. Commun.*, 1994, 991n/a.

85. A. K. Agrawal and S. A. Jenekhe, *Chem. Mater.*, 1993, **5**, 633.

86. S. A. Jenekhe, *Nature*, 1986, **322**, 345.

87. I. Osaka, G. Sauve, R. Zhang, T. Kowalewski and R. D. McCullough, *Adv. Mater.*, 2007, **19**, 4160.

88. Y. Zhu, R. D. Champion and S. A. Jenekhe, *Macromolecules*, 2006, **39**, 8712.

89. S. Ando, J. I. Nishida, H. Tada, Y. Inoue, S. Tokito and Y. Yamashita, *J. Am. Chem. Soc.*, 2005, **127**, 5336.

90. S. Ando, J. Nishida, E. Fujiwara, H. Tada, Y. Inoue, S. Tokito and Y. Yamashita, *Chem. Lett.*, 2004, **33**, 1170.

91. S. Ando, J. Nishida, Y. Inoue, S. Tokito and Y. Yamashita, *J. Mater. Chem.*, 2004, **14**, 1787.

92. S. A. Jenekhe, L. D. Lu and M. M. Alam, *Macromolecules*, 2001, **34**, 7315.

93. B. L. Lee and T. Yamamoto, *Macromolecules*, 1999, **32**, 1375.

94. M. Karikomi, C. Kitamura, S. Tanaka and Y. Yamashita, *J. Am. Chem. Soc.*, 1995, **117**, 6791.

95. W. Walker, B. Veldman, R. C. Chiechi, S. Patil, M. Bendikov and F. Wudl, *Macromolecules*, 2008, **41**, 7278.

96. J. Weber, M. Antonietti and A. Thomas, *Macromolecules*, 2008, **41**, 2880.

97. C. Yang, S. Cho, R. C. Chiechi, W. Walker, N. E. Coates, D. Moses, A. J. Heeger and F. Wudl, *J. Am. Chem. Soc.*, 2008, **130**, 16524.

98. M. Shkunov, D. Sparrowe, S. Tierney, R. Wagner, W. M. Zhang, M. L. Chabinyc, R. J. Kline, M. D. McGehee and M. F. Toney, *Nat. Mater.*, 2006, **5**, 328.

99. B. S. Ong, Y. L. Wu, P. Liu and S. Gardner, *J. Am. Chem. Soc.*, 2004, **126**, 3378.
100. J. Weber and A. Thomas, *J. Am. Chem. Soc.*, 2008, **130**, 6334.
101. N. Kleinhenz, L. Yang, H. Zhou, S. C. Price and W. You, *Macromolecules*, 2011, **44**, 872.
102. D. Gendron, P. O. Morin, P. Berrouard, N. Allard, B. R. Aich, C. N. Garon, Y. Tao and M. Leclerc, *Macromolecules*, 2011, **44**, 7188.
103. Y. Li, J. Zou, H. L. Yip, C. Z. Li, Y. Zhang, C. C. Chueh, J. Intemann, Y. Xu, P. W. Liang, Y. Chen and A. K. Y. Jen, *Macromolecules*, 2013, **46**, 5497.
104. P. Zhang, W. Li, X. Zhai, C. Liu, L. Daib and W. Liu, *Chem. Commun.*, 2012, **48**, 10431.
105. F. Bensebaa, A. A. Farah, D. Wang, C. Bock, X. Du, J. Kung and Y. L. Page, *J. Phys. Chem. B*, 2005, **109**, 15339.
106. Y. Haldorai, T. Zong and J. J. Shim, *Mater. Chem. Phys.*, 2011, **127**, 385.
107. K. H. Kate, K. Singh and P. K. Khanna, *Synth. React. Inorg., Met.-org., Nano-Met. Chem.*, 2011, **41**, 199.

CHAPTER 8

Polymer Modifications

8.1 Introduction

With the increase in research activities in the field of polymer science, the modification of polymers has acquired a key position in the current polymer situation. These modifications loosely come under the category of polymeranalogous reactions. Polymer modifications have proven to be a highly versatile tool-box for the functionalization and modification of macromolecules. The virtue of these reactions is that they aim at the tailor-made manipulation of polymer properties without changing the polymer backbone. Numerous strategies for polymeranalogous reactions involving microwave-assisted heating techniques have been planned and reported. The metal catalyzed "click" reaction of azides and alkynes has generated much buzz as a promising method to obtain complex polymer architectures with a high efficiency.[1,2] The conversion of other functional groups, for example, carboxylic or hydroxy groups by amidation or esterification, also provides suitable procedures for a successful polymer analogous reaction.

A great deal of work is being done on polymer modifications using microwave irradiations. Greater flexibility, improved product quality and properties, and the synthesis of new materials are some of the advantages that microwave irradiations provide which cannot be produced by other heating methods. These advantages have been detailed in several review articles on MW-assisted polymer processing. These papers highlight the recent advances in polymer modification including material fabrication, surface modification and synthesis of composites using microwave irradiation.[3–7] The application of microwave heating to polymers and polymeric composites has been pursued worldwide over the past 30 years. In microwave processing, energy is supplied by an electromagnetic field directly to the material, which results in rapid heating throughout the material thickness with reduced thermal gradients. Volumetric heating can also reduce

RSC Green Chemistry No. 35
Microwave-Assisted Polymerization
By Anuradha Mishra, Tanvi Vats and James H. Clark
© Anuradha Mishra, Tanvi Vats and James H. Clark 2016
Published by the Royal Society of Chemistry, www.rsc.org

processing times and save energy. The microwave field and the dielectric response of a material govern its ability to heat with microwave energy.[8] In conventional thermal processing, energy is transferred to the material through convection, conduction and radiation of heat from the mould surfaces, resulting in long cycle times and high-energy requirements. Thermal gradient during processing can result in uneven cure, residual stresses, and defects in the resultant composites. Polymer processing through microwaves offers many advantages over conventional thermal processing, however, the knowledge of electromagnetic theory and dielectric response is essential to optimize the processing of materials through microwave heating. Table 8.1 shows the SWOT (Strengths Weaknesses Opportunities and Threats) analysis of MW assistance in polymer processing.

Table 8.1   SWOT (Strengths Weaknesses Opportunities and Threats) analysis of MW assistance in polymer processing.

Strengths	Weaknesses
<ul><li>Cost savings (time and energy, reduced floor space)</li><li>Rapid heating of thermal insulators (most ceramics and polymers)</li><li>Precise and controlled heating (instantaneous on-off heating)</li><li>Selective heating</li><li>Volumetric and uniform heating (due to deep energy penetration)</li><li>Short processing times</li><li>Improved quality and properties</li><li>Synthesis of new materials</li><li>Processing not possible with conventional means</li><li>Reduction of hazardous emissions</li><li>Increased product yields</li><li>Environmentally friendly (clean and quiet)</li><li>Self-limiting heating in some materials</li><li>Power supply can be remote</li><li>Clean power and process conditions</li></ul>	<ul><li>Heating low-loss poorly absorbing materials</li><li>Controlling accelerated heating (thermal runaway)</li><li>Exploiting inverted temperature profiles</li><li>Eliminating arcing and controlling plasmas</li><li>Efficient transfer of microwave energy to work piece</li><li>Compatibility of the microwave process with the rest of the process line</li><li>Reluctance to abandon proven technologies</li><li>Timing</li></ul>
Opportunities	**Threats**
<ul><li>Development of compositions and processes tailored specifically for microwave processing</li><li>Better fundamental understanding and modeling of microwave/material interactions</li><li>Better process controls, electronic tuning and automation (smart processing)</li><li>More emphasis on microwave processing of magnetic materials</li></ul>	<ul><li>Availability of affordable equipment and supporting technologies</li><li>Availability of Kiln furniture, thermal insulation and other processing support hardware</li><li>Better communication among equipment manufacturers, technology developers, researchers and commercial users</li><li>Economics</li></ul>

In this chapter we have included the work done in the area of material processing using microwaves with special reference to polymer crosslinking and curing; polymer composites; polymer blends, and processing of polymeric scaffolds and particles.

8.2 Polymer Crosslinking/Curing/Derivitization

Curing is a process during which a chemical reaction like polymerization or a physical action like evaporation results in a harder, tougher or more stable linkage or substance. Some curing processes require maintenance of a certain temperature and/or humidity level, others require a certain pressure. Crosslinking or curing results in the formation of insoluble and infusible polymers wherein the polymer chains are joined together to give a three-dimensional structure. The process is pictorially represented in Figure 8.1 below.

The cure process at the molecular level involves several sequential stages. The first stage is the uncured stage, which consists of only a mixture of monomers or oligomers. On the commencement of the curing process, linear chain extension of the monomers and oligomers takes place. This is followed by chain crosslinking, which gives an effectively thermoplastic (flexible and moldable) phase. This process may take some time, and initially any rise on either the percent conversion or T_g may be detectable. Further heating would then result in chain crosslinking dominating the reaction to give a fully hardened solid (the commonly termed "C" stage) thermoset material, with significant shrinkage due to the molecular chains being pulled closer together by the crosslinks.[9]

Curing is of utmost importance in applications which require higher performance of thermoset polymers, such as adhesives and encapsulating sealant for the microelectronics industries. These requirements have led to unacceptably long cure times. In case of conventional thermal curing methods the cure rates increase, and in turn the cure times become lower.

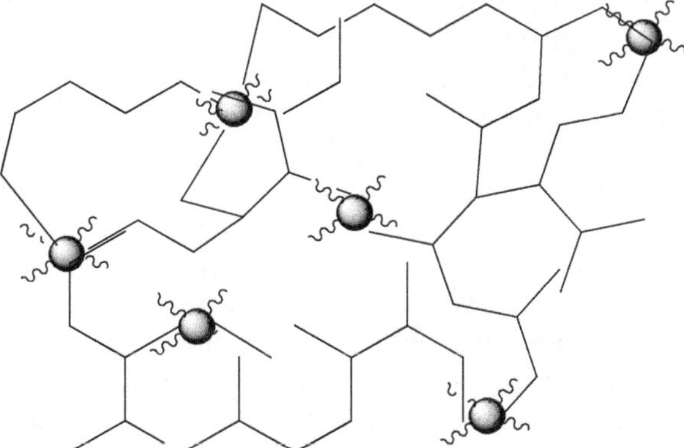

Figure 8.1 Diagrammatic representation of polymer crosslinking.

The major limitation of thermal curing is that the maximum rate of reaction for a given temperature is predetermined and heating beyond this limit would result in thermal degradation of the material. In order to overcome this limitation several methods like microwave curing, UV and electron beam curing have generated much interest.[10–24] The major difference between them is that microwave curing penetrates to deeper thicknesses, which is useful in polymer systems with a high percentage of high-density fillers. Microwave curing scores better for microelectronics applications where polymers are typically used with high filler contents. The fact that many filler/ceramic systems are actually transparent to microwaves makes microwave curing more important where the polymer resin is applied in areas optically inaccessible to UV or electron beam.

In order to fully comprehend and differentiate the thermal and microwave curing we need to obtain a temperature equivalent value. The curing process aims to obtain the maximum possible glass transition temperature in the shortest possible time. The glass transition temperature is usually determined using a percent conversion *versus* glass transition temperature relationship, such as the Di-Benedetto's equation,

$$T_g = T_{g0} + (T_{g\alpha} - T_{g0})\lambda x/[1 - (1 - \lambda)x] \tag{8.1}$$

where T_{g0} is the glass transition temperature of the uncrossed monomer and $T_{g\alpha}$ is the glass transition temperature of the fully reacted network and λ is an adjustable structural parameter. The reaction time can be determined by modeling the cure kinetic reaction. These methods, while reasonably established, are nonetheless based on thermal curing, the various parameters being functions of temperature. Microwave curing, on the other hand, has been best characterized as a function of electrical power (rate of energy).[10–14] It is thus essential to calculate a temperature equivalent value for microwave curing, which would then enable its results to be compared to the conventional thermal analysis methods.

Boey *et al.* wrote a comprehensive article on microwave irradiation curing of polymers. The article related the test results for both thermal and microwave curing by obtaining a temperature equivalent value using a phenomenological logarithmic approach. They showed that the equivalent temperature can be elucidated using a logarithmic plot of the cure times and the glass transition temperature. The values of the equivalent temperature so obtained were also consistently and significantly higher than the actual sample temperature during cure, which confirms that microwave curing follows a mechanism that is different to thermal curing.[9]

Guzeval and co-workers[25] presented experimental research which confirmed the benefits of the MW heating of curing polymer matrix for both laboratory and full-scale components over-heating in an electrical-heated oven. It was reported by the authors that MW cured polymers had greater load-bearing capacity and flexural modulus. Besides, the microstructure of the MW-cured polymer samples was uniform and the binder was distributed throughout the volume of polymer. Adhesive bonding in industrial production often requires very short curing times and high curing temperatures. The application of microwave radiation enables rapid processes and curing, even of bond lines that are thick or difficult to access.[26]

The advantages of variable frequency microwave curing of polymeric materials in terms of uniform curing in much less time, as compared to that obtained by conventional heating in microelectronics packaging applications, have also been investigated.[27–29]

Shimamoto *et al.*[30] in their study on the effects of microwave irradiation for resin-curing of carbon fiber/epoxy resin composite that was composed of discontinuous carbon fibers of 130 μm or 3 mm reported that the MW curing was nine times faster than curing by conventional heating. MW-cured carbon fiber/epoxy resin composite had higher glass transition temperature and much better mechanical properties as compared to that of the one prepared by a conventional oven. Zhou *et al.* used the advantages of microwave heating to study the curing of epoxy resin with different proportions of maleic anhydride as curing agent.[31] They compared the curing process using both microwave and thermal curing methods. The authors observed that the curing temperature of thermal treatment is about 15–20 °C higher than that of microwave treatment while the curing time is three times greater. They compared the compressive strength and bending strength of the samples and observed increased strength in samples cured by microwave treatment as compared to those cured by thermal treatment. The authors attributed their observations to a series of media polarization which takes place in epoxy resin in the presence of microwave irradiation. These include electronic polarization, atomic polarization, dipole steering polarization and interfacial polarization. These polarizations lead to an absorptive current forming the inner energy dissipation, which produces the inner heat in the epoxy resin, thus shortening the curing time. These processes collectively make the cured samples more homogeneous in structure, and do not overheat the surface of the cured samples. The authors also mentioned the fact that, because the frequency of the microwave is close to the revolving vibration frequency of chemical radicals, the conformation of the molecules could be changed. These dipole and electric field interactions lead the formation of a molecular net, and forms polymers with linear-structure or branched chain structure, which is favorable for improving the modulus of epoxy resin. Consequently, microwave curing improves the performances of polymer materials.[31] Improved mechanical properties of glass fibre epoxy composites prepared by resin transfer moulding and microwave curing have been reported.[32] Variable frequency microwave curing of amide-epoxy-based polymers and a comparison of microwave heating and conventional thermal heating for curing of carbon/epoxy composites has been reported by Tanaka *et al.* and Yusoff *et al.*[33,34]

Lee and co-workers studied the microwave process for the curing of waterborne polyurethane spin-coated on glass substrates.[35] The microwave cured sample showed an excellent caustic resistance compared to conventional cure, confirming that microwave heating produces a dense structure during the curing process. The dense structure does not affect the transmittance in the visible region. Bardts and Ritter described microwave-assisted synthesis of thiol modified polymethacrylic acid and its crosslinking with allyl modified polymethacrylic acid *via* thiolene "click" reaction. The thiol group was implemented by a polymer analogous condensation reaction of polymethacrylic acid and cysteamine, which was carried out in bulk by use of microwave. The allyl modified polymethacrylic acid was obtained by DCC-coupling of the allyl amine onto the polymethacrylic acid backbone.[36]

Microwave-assisted modifications and gelatinizations of polysaccharides, especially starch,[37] have been reported by many research groups. Interesting properties could be developed by MW treatment of polysaccharides. Microwave-assisted synthesis of tuneable mesoporous materials from α-*d*-Polysaccharides have been reported by White *et al.*[38] The enhancement of thermal, rheological and structural properties of cassava starch by microwave heating was reported by Colman *et al.*[39] Wheat starch gelatinization under microwave irradiation and conduction heating has been studied by Sainz *et al.*[40] They heated Wheat starch-water dispersions at excess water conditions under mixing to different temperatures by either microwave energy at 2000 W or by conduction heating in order to compare both heating methods. The effects of microwaves and conduction heating on starch gelatinization were evaluated using different techniques. The authors obtained the same viscosity for both heating methods but the required time was much shorter for microwaves heated starch dispersions. No unique structures due to the heating method were observed under the microscope, indicating that the mechanism of gelatinization remained unaffected by the heating method. DSC experiments revealed an increase in the enthalpy for short times with microwave-heated samples. X-ray experiments showed that the heating method did not alter the gelatinization mechanism; however, the rate of loss of crystallinity with temperature was higher for microwave-heated samples, which was attributed to a more complete gelatinization, as revealed by the NMR results. Physico-chemical changes induced in a wheat starch model system as a result of microwave heating also suggested a different mechanism of starch gelatinization compared to conduction heating. Starch granule rupture and formation of film polymers coating the granule surface was also reported as the consequence of vibrational motion and the rapid increase in temperature during MW heating.[41] The results of microwave treatment of potato starch granules demonstrated that the morphology and crystalline structure was damaged by microwave treatment. The fracture of starch granules, molecular chains leached from the granules and degradation of dextran chains contributed to the development of rheological properties.[42]

The effect of microwaves on chemical modifications of starch has also been studied extensively.[43] Microwave-assisted oxidation of starch using hydrogen peroxide as an oxidant presented significant differences in the degree of oxidation (monitored by means of carboxyl and carbonyl group content) as well as the changes in molecular mass of polysaccharide comparing to those observed at conventional conditions.[44] The kinetics of starch oxidation using the same oxidant has also been studied by another group.[45]

Esterification of starch expands the usefulness of starch for a myriad of industrial applications.[46] Enzymatically catalyzed esterification of starch using low power microwave heating and solvent-free lipase-catalyzed synthesis of long-chain starch esters under MW irradiation were reported.[47,48] The hydrophobic starch fatty acid esters produced may have potential industrial applications such as surface-coating materials, flavoring agents in the food industry and biomedicals for bone fixation and replacements.[47] The experimental results showed that the degree of substitution was found to be higher in MW esterification than in liquid state esterification.[46] The esterification of starch with fatty acids enhances its thermoplastic character and its mechanical properties, increases its thermal stability and renders it hydrophobic. These improved properties are higher when the carbon chain length is longer and the degree of substitution of starch esters is higher. Additionally, the tendency of starch to swell in water is eliminated by esterification. These properties are of great advantage in the plastics and pharmaceutical industries and in biomedical applications such as materials for bone fixation and replacements, carriers for controlled release of drugs and other bioactive agents.[48,49] Rapid homogeneous esterification of cellulose by microwave irradiation has been reported by Chadlia and Farouk.[48]

Microwave-assisted boration[50] and carboxymethylation[51] and accelerated methylation[52] of starch have also been reported. A recent review on microwave-assisted polycondensation reactions and curing of products thus obtained has been published by Komorowska-Durka *et al.*[53]

8.3 Polymer Composites

A composite material is a combined material created from two or more components, to obtain specific characteristics or properties that were not there before. The matrix is the continuous phase, and the reinforcement constitutes the dispersed phase. It is the behavior and properties of the interface that generally control the properties of the composite. A hybrid is a composite which usually combines an organic and inorganic material. The difference between the hybrids and composites lies in the scale of mixing. In "hybrid materials" the level of mixing of the different types of materials is at the nanometer level, or sometimes at the molecular level.[54,55] Figure 8.2

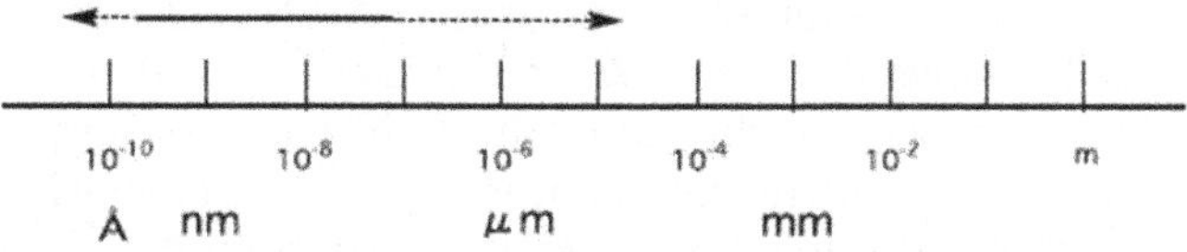

Figure 8.2 Domain size in hybrid materials.

gives the size scale, where the continuous line part in the figure is the hybrid materials region.

The use of microwave heating to create polymers and polymeric composites has been pursued worldwide over the past 30 years. Rapid volumetric heating in microwave processing significantly reduces cycle times and, additionally, since power is applied directly to the material, the need to heat processing equipment is avoided, further reducing the energy requirement. MW processing results in even cure with no residual stresses or defects in the resultant composites.

The reinforcement can be polymer and carbon fibers and particles and glass or ceramic fillers, while the matrix is usually of polymeric nature. Polymer composites have several specific applications in the fields of electricals, electronics, microelectronics, marine technology and aerospace. The ability of the matrix to transfer an applied stress to the reinforcement results in improved mechanical characteristics, appropriate for fulfilling structural and mechanical roles in hard tissue repair and dentistry. The energy transfer process relies on intimate reinforcement/matrix interaction. Thus, the better the interaction attained, the greater the mechanical properties of the material.[56]

Papargyris *et al.* compared the mechanical properties of carbon fiber/epoxy composites produced by the conventional method and MW.[57] On exposure to four-point bending and interlaminar shear testing, samples produced by the conventional method exhibited poorer interfacial bonding and greater extensive fiber pullout than the composites produced by MW irradiation. The authors attributed this improvement of the interlaminar shear strength for MW-produced samples to the lowering of the matrix viscosity at the initial stages of the curing process due to more effective heating. This phenomenon facilitated better resin flow and interfacial resin-fiber bonding.[58]

Fullerene (Figure 8.3) has become a central research attraction owing to its unique electronic conducting, magnetic and photophysical properties. Poor solubility and difficulty in processing are the two factors that act as blocks in direct application. In order to overcome these challenges, grafting polymer chains on fullerene has become as a viable option. These fullerene-based specialty polymeric materials have peculiar physicochemical properties and good processability. Therefore, several synthetic strategies have been developed for preparing soluble fullerene-containing polymers, and different kinds of fullerene functionalized polymers.[59–61]

Wu and co-workers performed fullerenation of polycarbonate (PC), a commercially important optical polymer, by direct reaction of C_{60} and PC in the presence of azo-bis-isobutyronitrile (AIBN), using 1,1,2,2-tetrachloroethane as the solvent under MW irradiation.[62] The major advantage of the C_{60}-PC composites is that they are soluble in common organic solvents such as THF and chloroform. On comparing with conventional heating process, the authors observed that MW irradiation could significantly enhance the rate of the fullerenation under identical reaction conditions.

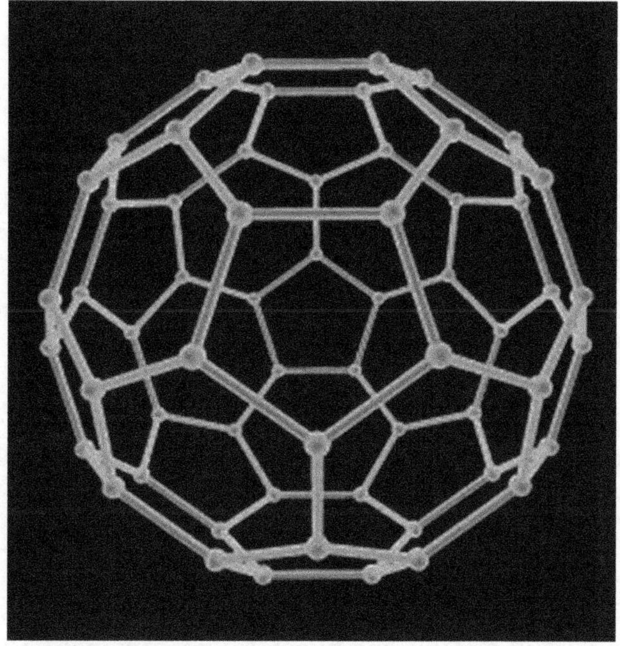

Figure 8.3 Buckminsterfullerene C_{60}.

The authors monitored C_{60} content of the fullerene functionalized polycarbonate (C_{60}-PC) by varying the C_{60}/PC feed ratio and the reaction time and characterized the products by gel permeation chromatography, UV–vis, FTIR, TGA, DSC, ^{1}H NMR and ^{13}C NMR. The electron spin resonance spectra indicated the presence of the fullerene radicals in reaction solutions and also in the solid product polymers, thus proving the radical mechanism of the reaction. The nonlinear optical property of C_{60}-PCs in THF was investigated by the open-aperture z-scan technique at 527 nm, and its nonlinear absorption coefficient was found to be in the same order as that of C_{60}. With the improved solubility, the C_{60}-PCs synthesized under MW irradiation could potentially be applied as effective nonlinear optical materials.

The C_{60} end-capped polymers were obtained with the bromo double-terminated polymers were obtained from the ATRP of styrene and MMA using α,α-dibromo-p-xylene under MW irradiation.[63] The precursor bromo-terminated polymers were subsequently functionalized with fullerene C_{60} using CuBr/bipy as the catalyst system under MW irradiation. The reaction is depicted in the following scheme (Scheme 8.1).

The telechelic C_{60} end capped products were characterized by the similar techniques as used for C_{60}-PC composites. From the results, it could be concluded that microwave irradiations significantly enhanced the rate of fullerenation reaction, and the physical properties and structure of the C_{60} endcapped polymers are not modified by the use of the microwave.

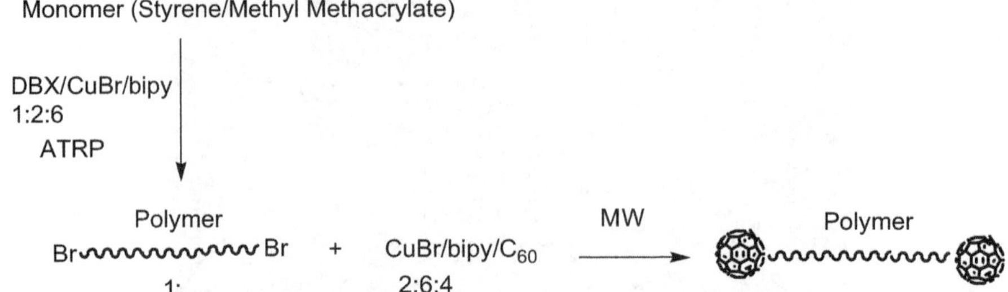

Scheme 8.1   MW-assisted ATRP of Styrene and methyl Methacrylate using α,α-dibromo-*p*-xylene to obtain C$_{60}$ end capped polymers.[44]

Nadagouda and Varma devised a facile method utilizing MW irradiation that utilizes the crosslinking reaction of polyvinyl alcohol (PVA) with metallic and bimetallic systems.[64] The authors prepared nanocomposites of PVA crosslinked metallic systems such as Pt, Cu and In and bimetallic systems such as Pt–In, Ag–Pt, Pt–Fe, Cu–Pd, Pt–Pd and Pd–Fe by reacting the respective metal salts with 3 wt% PVA under microwave irradiation, maintaining the temperature at 100 °C. The procedure provides a simple route to manufacturing useful metallic and bimetallic nanocomposites with various shapes, such as nanospheres, nanodendrites and nanocubes. The authors stressed the advantages of bimetallic catalyst systems,[64,65] which included control over the catalytic activity, selectivity and stability, and some combinations may exhibit synergistic effects.[65–69] All these factors led to the improved "catalyst atom economy".[70,71] The authors maintained that the constant temperature employed in these studies does not establish a correlation of the particle size with the MW power applied, adding that more exploratory studies are required to understand the particle-growth mechanism. Going by the same method, the authors reported the crosslinking reaction of PVA with single-walled carbon nanotubes (SWNTs), multi-walled carbon nanotubes (MWNTs) and C$_{60}$ using MW irradiation.[72] The authors claimed a radical improvement over literature methods in preparation of nanocomposites of PVA crosslinked with SWNT, MWNT and C$_{60}$. The composites were prepared expeditiously by reacting the respective carbon nanotubes with 3 wt% PVA under MW irradiation, maintaining a temperature of 100 °C. Enhanced coupling of SWNT with MW was observed, when compared to related MWNT and C$_{60}$, which resulted in better crosslinking. The general MW-assisted synthesis proved to be versatile, green in nature and provided a simple route to manufacture useful SWNT, MWNT and C$_{60}$ PVA nanocomposites. Carbon nanotubes (CNTs) have been shown to absorb microwave radiation and readily convert it to heat with extremely high efficiency. Microwave-assisted cellulose-based nanocomposites and derivatives have recently been published by Ming-Guo.[73]

Microwave-assisted healing of thermally mendable composites were reported by Sosa *et al.*[74,75] Self-healing materials[76] represent the latest area of interest in materials chemistry and engineering. Microwave-assisted self-healing materials, such as four-arm star polymers based on double network formation due to "click" crosslinking by MW and supramolecular cluster formation has been reported by Dohler *et al.*[77]

Tang *et al.* synthesized organic-inorganic polyacrylamide-calcium phosphate nanocomposites from calcium chloride, ammonium phosphate and acrylamide monomer in aqueous solution by a single-step microwave-assisted method.[78] Kajiwara *et al.* used sol–gel process to synthesize a poly(2-hydroxyethyl methacrylate) (PHEMA)/silica hybrid.[79] It was observed that, though the reaction rate was increased by MW irradiation, the thermal properties remained almost unchanged with respect to the same material produced under conventional heating. Additional studies explored the effect of a post-synthesis MW-heating on the mechanical properties of the material. Data consistently indicated that the post treatment results in greater strength and modulus.[80,81]

Ma and co-workers synthesized electrically conductive polymer composites, glycerol plasticized-starch (GPS)/carbon black (CB) membranes by melt extrusion and MW radiation.[82] The comparative report mainly comprised characterization of the composites formed by the two methods. With the aid of scanning electron microscopy graphs it was established that the electrical conductance network of CB is formed in GPS/CB membranes, prepared by microwave radiation (GPS/CB-MW). Whereas the agglomerates of CB particles are observed in GPS/CB membranes, prepared by melt extrusion (GPS/CB-ME). FTIR spectroscopy revealed that CB and GPS matrix can form the interaction in GPS/CB membranes. Interfacial interaction between the fillers and matrix is an important factor affecting the mechanical properties of the composites. According to the Nicholais–Narkis models, the reinforcing effect of CB is more obvious in GPS/CB-MW membranes than in GPS/CB-ME membranes. GPS/CB-MW membranes exhibit a low electrical percolation threshold of 2.398 vol% CB loading and the conductivity of the membrane containing 4.236 vol% CB reaches 7.08 S/cm, while GPS/CB-ME membranes shows a very low conductivity of 10^{-8} S/cm at the high CB content. Also, GPS/CB-MW membranes reportedly have a better water vapor barrier than GPS/CB-ME membranes.

Adeodu *et al.*[83] reported the effect of microwave post curing on the cure cycles of the unsaturated polyester composites reinforced with aluminum and carbon black. They reported that post-curing of the particulate composites through microwaving is able to achieve a good heating rate and better control of temperature as compared to the conventional curing. Microwave heating in the post curing of polymer matrix composites not only reduces the lengthy cure cycle but also improves the mechanical properties of the composite produced.

The MW-assisted ring-opening polymerization of lactones utilizing surface-functionalized inorganic and organic reinforcing agents such as hydroxyapatite,[84] carbon nanotubes,[58,85] clays[85,86] and nanocrystalline starch[87] is a common synthetic method to obtain biomedical composites.

8.4 Polymer Scaffolds

The MW synthesized/assisted polymeric materials are very useful in the biomedical research arena. In general, better interfacial integration and greater mechanical performances were observed under MW irradiations.

Cells are often implanted or "seeded" into an artificial structure capable of supporting three-dimensional (3D) tissue formation. These 3-D structures are called scaffolds. Scaffolds have been recognized since the mid-1970s as essential components for the development of engineered tissues and organs. The development of artificial extracellular matrices to serve as templates for cell attachment/suspension, proliferation, growing and delivery has progressed at a tremendous rate in recent years, which in turn has fueled research activity into the science of scaffolds. Scaffolds usually aim to serve at least one of the following purposes.[20,88,89]

✓ Allow cell attachment and migration.
✓ Deliver and retain cells and biochemical factors.
✓ Enable diffusion of vital cell nutrients and expressed products.
✓ Exert certain mechanical and biological influences to modify the behavior of the cell phase.

The scaffold construct supports cells to proliferate and maintain their differentiated function, and its architecture defines the ultimate shape of the new bone and cartilage. Several scaffold materials have been investigated for tissue engineering bone and cartilage, including hydroxyapatite (HA), poly(a-hydroxyesters) and natural polymers such as collagen and chitin. Many hydrogels, as discussed in Chapter 6, make scaffolds. In order to fully comprehend the function, properties and rational of application of polymer-based scaffolds in tissue engineering, it is important to discuss the terms "biodegradable", "bioerodable", "bioresorbable" and "bioabsorbable", which find a repeated reference in this field.[20]

Biodegradables are solid polymeric materials and devices which break down due to macromolecular degradation with dispersion *in vivo* but there is no proof of the elimination from the body (this definition excludes environmental, fungi or bacterial degradation). Biodegradable polymeric systems or devices can be attacked by biological elements so that the integrity of the system, and in some cases but not necessarily, of the macromolecules themselves, is affected and gives fragments or other degradation byproducts. Such fragments can move away from their site of action but not necessarily from the body.

Bioresorbables are solid polymeric materials and devices which show bulk degradation and further resorb *in vivo*; *i.e.* polymers which are eliminated through natural pathways either because of simple filtration of degradation byproducts or after their metabolization. Bioresorption is thus a concept which reflects total elimination of the initial foreign material and of bulk degradation of byproducts (low molecular weight compounds) with no residual side effects. The use of the word "bioresorbable" assumes that elimination is shown conclusively.

Bioerodibles are solid polymeric materials or devices which show surface degradation and further resorb *in vivo*. Bioerosion is thus a concept, too, which reflects total elimination of the initial foreign material and of surface degradation byproducts (low molecular weight compounds) with no residual side effects.

Bioabsorbables are solid polymeric materials or devices which can dissolve in body fluids without any polymer chain cleavage or molecular mass decrease. For example, it is the case of slow dissolution of water-soluble implants in body fluids. A bioabsorbable polymer can be bioresorbable if the dispersed macromolecules are excreted.

These definitions are given by Vert,[90] and fully explain and correctly define the often misinterpreted words in tissue engineering. Many hydrogels described in Chapter 6 form scaffolds. Given below is the step-wise procedure involved in tissue scaffolding (Figure 8.4).

The polymeric scaffolds have been recognized since the mid-1970s as key components for the development of engineered tissues and organs. Over time, the development of artificial extracellular matrices to serve as templates for cell attachment/suspension, proliferation, growing and delivery has gained a lot of momentum. A plethora of methods are now available for

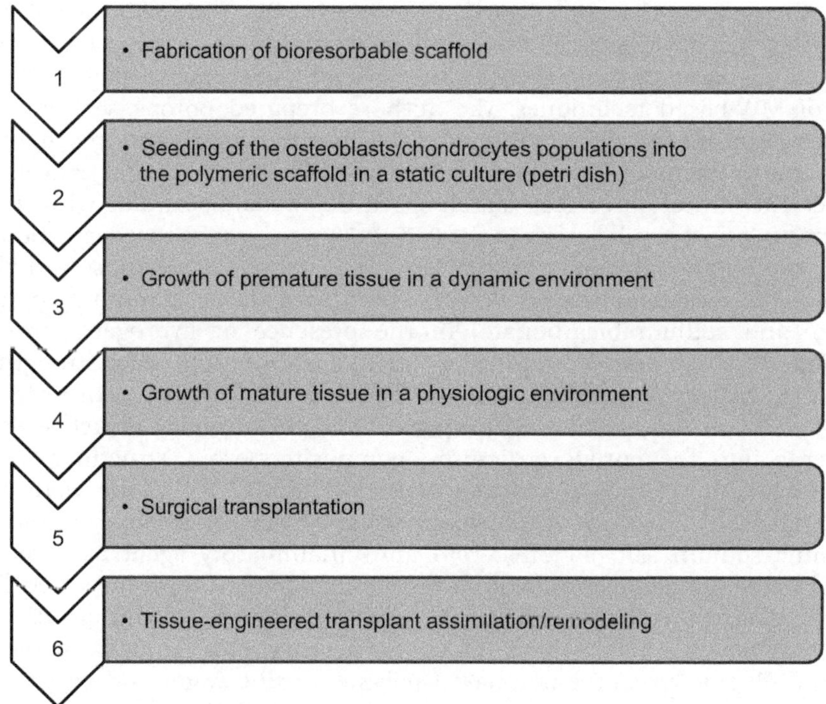

Figure 8.4 Step-wise procedure of tissue scaffolding.

developing 3D porous scaffolds using polymeric materials for tissue-engineering applications. Amongst them, solvent casting and particulate leaching, membrane lamination, fiber bonding, phase separation/inversion, melt-based technologies, microparticle aggregation and MW baking and expansion, have proved to be highly successful. Zein and co-workers developed fused deposition modeling to produce novel scaffolds with a honeycomb-like pattern, a fully interconnected channel network, having controllable porosity and channel size.[91] The authors developed a bioresorbable polymer poly(ε-caprolactone) (PCL), which was used as a filament modeling material to produce porous scaffolds, made of layers of directionally aligned microfilaments. Depending on the processing parameters, the PCL scaffolds produced had a range of channel size 160–700 mm, filament diameter 260–370 mm and porosity 48–77%, and possessed regular geometrical honeycomb pores. The scaffolds of different porosity also exhibited a pattern of compressive stress-strain behavior characteristic of porous solids under such loading. The compressive stiffness ranged from 4 to 77 MPa, yield strength from 0.4 to 3.6MPa and yield strain from 4% to 28%. Lam used a similar technique for scaffold development using 3D printing with a starch-based polymer.[92] Many of these technologies fail to provide the scaffolds with basic requirements such as pore interconnectivity and suitable mechanical properties, and most of them are only able to produce scaffolds with specific requirements for a limited number of applications.

It was in 2001 that Malafaya *et al.* reported MW processing of polymer scaffolds[93] as an innovative methodology for producing porous biodegradable starch-based tri-dimensional structures for tissue engineering. With the aid of MW-based techniques, the authors prepared porous scaffolds exhibiting an interesting combination of morphological and mechanical properties very much like the compressive behavior of human cancellous bone. The authors produced starch-based degradable scaffolds with interconnecting pores and density in the 0.40–0.50 g/cm^3 range that may find use in tissue engineering or drug delivery applications. It was observed that formulations containing 10% blowing-agent (corn starch, sodium pyrophosphate and sodium bicarbonate) in the presence of hydrogen peroxide exhibited a compression modulus of 530 MPa and a compressive strength of 60 MPa, values close to those of the native cancellous bone. In order to improve upon the mechanical properties and to introduce bioactive components into the porous structures, composite porous structures using hydroxyapatite as filler by means of a similar MW technique were also investigated.[94] Loaded drug delivery porous carriers, incorporating meclofenamic sodium salt, a non-steroid anti-inflammatory agent as a model drug, were obtained using a similar processing strategy. It was expected that the developed methodology might be used for other drugs and growth factors that play a crucial role in tissue engineering. Biodegradable scaffolds from different types of starch-based polymers using water and glycerol as plasticizers were also prepared under MW irradiations.[95] Compressive mechanical properties were assessed by indentation tests, and a strong dependence of the indentation stress on the average pore size was found. Another methodology reported for the fabrication of 3D porous scaffolds was MW radiation under vacuum.[96] MW-assisted plasma has also been intended for the post treatment surface modification of scaffolds with greater efficiency than the standard plasma technique.[97] MW applies higher frequency radiation. The benefits of this approach have already been apparent for improving the chondrogenic response of porous silk fibroin scaffolds.[98]

Besides hydrogel synthesis, MW-assisted synthesis has slowly made its presence felt in ionogel synthesis as well. Ionogels are solid electrolytes comprised mainly of nonvolatile, room temperature ionic liquids with a small amount of a solid supporting scaffold material.[99] In a recent study, MW-assisted ionogl synthesis has been reported. The basic idea behind the synthesis was the fact that the ability of a solvent to absorb energy *via* microwave irradiation allows for a swift temperature rise in ionic liquid that can be enhanced for solvents that exhibit ionic conductivity.[100] Visentin *et al.* reported rapid, microwave-assisted thermal polymerization of poly(ethylene glycol) diacrylate-supported ionogels.[101] The major advantage of this method over commonly employed methods like the use of a co-solvent to blend in a polymer support,[102] initiation of a spontaneous sol–gel reaction to produce an inorganic support,[103] stirring in an assembly of fumed silica particles,[104] and either UV-initiated[105] or thermally initiated[106] polymerization/cross-linking of a reactive monomer inside the ionic liquid is the reduced reaction times (seconds as compared to days). The ionogels hence produced exhibited electrical and mechanical properties that are comparable to those of ionogels fabricated by other methods. Post this ground-breaking development, further control of the MW-assisted heating profile may lead to even faster fabrication times for high-performance, solid ionogel electrolytes.

Zhihong and Yao[107] reported that the shape memory effects of an originally thermally activated shape memory polymer (SMP) matrix may be initiated by a touchless and highly selective microwave stimulus by the incorporation of T-ZnOw particles in SMP. Microwave heating as a novel method has been used by Zhang *et al.*[108] to rapidly foam and actuate biocompatible and biodegradable shape memory crosslinked-polycaprolactone (c-PCL) foams. These c-PCL foams have homogenous porosity and their shape recovery ratio was approximately 100% and the whole recovery process took only 98 s when trigged by microwave.

Shape memory polymers are an interesting class of smart materials.[109] Using microwave irradiation for their quick synthesis needs to be explored.

8.5 Miscellaneous

Spasojevic and his research team compiled an extensive study of the effects of conventional isothermal heating and microwave isothermal heating (CIH and MWIH) on the properties including residual monomer, hardness and water absorption of three commercial poly(methyl methacrylate) (PMMA) base dental materials.[110] It was observed that the rate of the MWIH polymerization was eight times higher compared with the CIH polymerization. In case of the CIH polymerization the mechanism appeared to be the phase-boundary controlled reaction for which the rate of contracting volume is the rate-limiting step, whereas in MWIH polymerization, the mechanism was first-order reaction and the monomer concentration in the polymerization mixture was the rate-limiting step. The results established that the samples of PMMA base dental materials synthesized by MWIH polymerization exhibit better hardness and water absorption properties. The hardness of a material is increased, while the content of the residual monomer and water absorption is decreased in the MWIH compared with the CIH polymerized samples for the same polymerization temperature and time.

Microwave heating has been used extensively for the modification of starch/cellulosic materials. Synthesis of cellulosic plastic films, obtained in homogeneous conditions, by microwave-induced acylation of commercial or chestnut tree sawdust cellulose by fatty acids has been studied by Joly *et al.*[111] The acylation reaction was quantified according to *N,N*-dimethyl-4-aminopyridine (DMAP) amount where DMAP simultaneously acted as catalyst and proton trapping base. The study indicated that DMAP does not influence degrees of substitution, massic and molar yields. Plastic films synthesized in the absence of DMAP showed a decrease in mechanical behavior. Microbial biodegradation of plastic films with DS = 2.2 led to a loss of their mechanical behaviors.

Ebringerova and co-workers[112] reported microwave-assisted synthesis of carboxymethylcellulose based polymeric surfactants. Carboxymethylcellulose (CMC, DSCM = 1) was partially hydrophobized in order to prepare polymeric surfactants by the transesterification reaction using the methyl ester of the fatty acid complex of rapeseed oil (MERO). The chemical modification was performed in different reaction media (i) DMF/TSA and (ii) H_2O/DMF with and without K_2CO_3 as catalyst, at various reaction conditions and using MW radiation with controlled power as a heating source. The FTIR and NMR analysis of MERO-hydrophobized CMC (MH-CMC) comprising mixed fatty acyl esters indicated a very low degree of esterification (DS < 0.1). Irrespective of moderate surface tension-lowering effects, the derivatives exhibited excellent emulsifying activity for "oil in water" type emulsions as well as good washing power and antiredeposition efficiency. The observations indicted that MW heating reduced the synthesis time of surface-active MH-CMC to a few minutes compared to 6 h under conventional heating. The transesterification method could be a potential substitute for the toxic, hazardous and time-consuming classical esterification processes in preparing polysaccharide-based surfactants. The proven applicability of MERO as acylation agent contributes to a further valorization of the rapeseed biomass. The novel MW-assisted CMC esters represent biodegradable polymeric surfactants with potential applications in manufacture of consumer products and in industrial processes.

Microwave-assisted modification of starch has been done for compatibilizing LLDPE/starch blends.[113] Cassava starch was successfully modified with octenyl succinic anhydride (OSA) by the aid of a MW-assisted reaction in solid state under alkaline conditions. In comparison to native cassava starch, the modified starches exhibited less hydrophilicity, which in turn led to their application as compatibilizers in LLDPE/starch blends. An increase in Young modulus observed in the LLDPE/starch blends was attributed to the rigid character of the filler. A poor interfacial adhesion was cited the reason for the reduction in tensile strength with increasing starch content. However, the mechanical properties were significantly improved by adding 10% of OSA with DS = 0.045. The blends compatibilized with OSA under MW irradiations exhibited the stress at yield is basically the same as that of neat LLDPE.

Polymer modifications using MW is still not a fully blossomed research field, even after a lot of work has been done and reported. Microwave heating is expected to bring more advantages to modern industry and scientific research in the field of smart materials and structures.[114] A lot more is to be done to upgrade the process to an industrial scale.

References

1. R. M. Paulus, T. Erdmenger, C. R. Becer, R. Hoogenboom and U. S. Schubert, *Macromol. Rapid Commun.*, 2007, **28**, 484.
2. C. Guerrero-Sanchez, R. Hoogenboom and U. S. Schubert, *Chem. Commun.*, 2006, **36**, 3797.
3. D. E. Clark and W. H. Sutton, *Annu. Rev. Mater. Sci.*, 1996, **26**, 299.
4. L. Zong, S. Zhou, N. Sgriccia, M. C. Hawley and L. C. Kepel, *J. Microwave Power EE*, 2003, **38**, 49.
5. B. Dariusz, P. Piotr, P. Jan and P. Aleksander, *Adv. Polym. Sci.*, 2003, **163**, 193.
6. D. Bogdal and A. Prociak, *Chim. Oggi.*, 2007, **25**, 30.
7. S. J. Lee, K. S. Sandhu and S. T. Lim, *J. Food Sci. Biotechol.*, 2007, **16**, 832.
8. E. T. Thostenson and T. W. Chou, *Composites, Part A*, 1999, **30**, 1055.
9. F. Y. C. Boey and S. K. Rath, *Adv. Polym. Technol.*, 2000, **19**, 194.
10. F. Y. C. Boey, B. F. Yap and L. Chia, *Polym. Test.*, 1999, **20**, 837.
11. J. Zhou, C. Shi, B. Mei and R. Yuan, *J. Mater. Process. Technol.*, 1993, **137**, 156.
12. C. O. Kappe, *Angew. Chem.*, 2004, **43**, 6250.
13. J. Jacob, L. H. L. Chia and F. Y. C. Boey, *J. Appl. Polym. Sci.*, 1996, **63**, 787.
14. L. H. L. Chia, F. Y. C. Boey and J. Jacobs, *J. Polym. Sci.*, 1996, **34**, 2087.
15. F. Y. C. Boey, *Polym. Test.*, 1995, **14**, 471.
16. J. Jacob, L. H. L. Chia and F. Y. C. Boey, *J. Mater. Sci.*, 1995, **30**, 5321.
17. F. Y. C. Boey and C. Y. Yue, *J. Mater. Sci. Lett.*, 1991, **10**, 1333.
18. J. Wei, M. C. Hawley, J. D. Delong and M. Demeuse, *Polym. Eng. Sci.*, 1993, **33**, 1132.
19. E. Marand, K. R. Baker and J. D. Graybeal, *Macromolecules*, 1992, **25**, 2243.

20. D. A. Lewis, J. D. Summers, T. C. Ward and J. E. McGrath, *J. Polym. Sci.*, 1992, **30**, 1647.
21. J. C. Hedrick, D. A. Lewis, G. D. Lyle, T. C. Ward and J. E. McGrath, *Polym. Prep.*, 1988, **29**, 363.
22. M. Delmotte, H. Jullien and M. Ollivon, *Eur. Polym. J.*, 1991, **27**, 371.
23. E. Marand, K. R. Baker and J. D. Graybeal, *Polym. Prep.*, 1990, **31**, 397.
24. D. A. Lewis, J. C. Hedrick, J. E. McGrath and T. C. Ward, *Polym. Prep.*, 1987, **28**, 330.
25. T. Guzeval, Y. Zavitaeva and E. Chutskova, *IOP Conf. Series: Mater. Sci. Eng.*, 2015, **74**, 012007.
26. A. Maurer and C. Lammel, *Adhesion Adhesives Sealants*, 2014, **11**, 26.
27. S. K. Pavuluri, G. Goussetis, M. P. Y. Desmulliez, T. Tilford, R. Adamietz, G. Muller, F. Eicher and C. Bailey, *IEEE Trans. Compon., Packag. Manuf. Technol.*, 2012, **2**, 799.
28. S. Qi, B. Vaidhyanathan and D. Hutt, *J. Mater. Sci.*, 2013, **48**, 7204.
29. M. D. Diop, M. Paquet, D. Danovitch and D. Drouin, *IEEE Trans. Device Mater. Reliab.*, 2015, **15**, 250.
30. D. Shimamoto, Y. Imai and Y. Hotta, *OJCM*, 2014, **4**, 85.
31. J. Zhou, C. Shi, B. Mei, R. Yuan and Z. Fu, *J. Mater. Process. Technol.*, 2003, **137**, 156.
32. C. Xu, Y. Gu, Z. Yang, Min Li, Y. Li and Z. Zhang, *J. Compos. Mater.*, in press, DOI:10.1177/0021998315590259.
33. K. Tanaka, S. A. B. Allen and P. A. Kohl, *IEEE Trans. Compon. Packag. Technol.*, 2007, **30**, 472.
34. R. Yusoff, M. K. Aroua, A. Nesbitt and R. J. Day, *J. Eng. Sci. Technol.*, 2007, **2**, 151.
35. H. Lee, C. Y. Fang, C. G. Pantano and W. Kang, *Bull. Korean Chem. Soc.*, 2011, **32**, 1.
36. M. Bardts and H. Ritter, *Macromol. Chem. Phys.*, 2010, **211**, 778.
37. M. Brasoveanu and M. R. Nemtanu, *Starch-Stärke*, 2014, **66**, 3.
38. R. J. White, V. Budarin, R. Luque, J. H. Clark and D. J. Macquarrie, *Chem. Soc. Rev.*, 2009, **38**, 3401.
39. T. A. D. Colman, I. M. Demiate and E. Schnitzler, *J. Therm. Anal. Calorim.*, 2014, **115**, 2245.
40. C. Bilbao-Sainz, M. Butler, T. Weaver and J. Bent, *Carbohydr. Polym.*, 2007, **69**, 224.
41. T. Palav and K. Seetharaman, *Carbohydr. Polym.*, 2007, **67**, 596.
42. Y. Xie, M. Yan, S. Yuan, S. Sun and Q. Huo, *Chem. Cent. J.*, 2013, **7**, 113.
43. K. Lewicka, P. Siemion and P. Kurcok, *Int. J. Polym. Sci.*, 2015, 867697, DOI: 10.1155/2015/867697.
44. M. Lukasiewicz, S. Bednarz and A. Ptaszek, *Starch-Stärke*, 2011, **63**, 268.
45. P. Tolvanen, P. Mäki-Arvela, A. B. Sorokin, T. Salmi and D. Y. Murzin, *Chem. Eng. J.*, 2009, **154**, 52.
46. A. Rajan, J. Sudha and T. E. Abraham, *Ind. Crop. Prod.*, 2008, **27**, 50.
47. M. Lukasiewicz and S. Kowalski, *Starch-Stärke*, 2012, **64**, 188.
48. A. Chadlia and M. M. Farouk, *J. Appl. Polym. Sci.*, 2011, **119**, 3372.

49. H. Horchani, M. Chaabouni, Y. Gargouri and A. Sayari, *Carbohydr. Polym.*, 2010, **79**, 466.

50. H. Staroszczyk, *Polymers*, 2009, **54**, 31.

51. J. Liu, J. Ming, W. J. Li and G. H. Zhao, *Food Chem.*, 2012, **133**, 1196.

52. V. Singh and A. Tiwari, *Carbohydr. Res.*, 2008, **343**, 151.

53. M. Komorowska-Durka, G. Dimitrakis, D. Bogdai, A. I. Stankiewicz and G. D. Stefanidis, *Chem. Eng. J.*, 2015, **264**, 633.

54. Y. Chujo, *Micrometrics*, 2007, **50**, 11.

55. A. H. Yuwono, J. Xue, J. Wang, H. I. Elim, W. Ji, Y. Lic and T. J. White, *J. Mater. Chem.*, 2003, **13**, 1475.

56. T. Sato, K. Masaki, K. Sato, Y. Fujishiro and A. Okuwaki, *J. Chem. Technol. Biotechnol.*, 1996, **67**, 339.

57. D. A. Papargyris, R. J. Day, A. Nesbitt and D. Bakavos, *Compos. Sci. Technol.*, 2009, **68**, 1854.

58. F. Buffa, H. Hu and D. E. Resasco, *Macromolecules*, 2005, **38**, 8258.

59. L. Dai, *Polym. Adv. Technol.*, 1999, **10**, 357.

60. L. Dai and W. H. A. Mua, *Adv. Mater.*, 2001, **13**, 899.

61. K. E. Geckeler and S. Samal, *Polym. Int.*, 1999, **48**, 743.

62. H. Wu, F. Li, Y. Lin, R. Cai, H. Wu, R. Tong and S. Qian, *Polym. Eng. Sci.*, 2006, **46**, 399.

63. H. Wu, F. Li, Y. Lin, M. Yang, W. Chen and R. Cai, *J. Appl. Polym. Sci.*, 2006, **99**, 828.

64. M. N. Nadagouda and R. S. Varma, *Macromol. Rapid Commun.*, 2007, **28**, 465.

65. L. Guczi, *Catal. Today*, 2005, **101**, 53.

66. M. Takanori and T. Asakawa, *Appl. Catal. A*, 2005, **280**, 47.

67. M. T. Reetz, M. Winter, R. Breinbauer, T. Thurn-Albrecht and W. Vogel, *Chem.– Eur. J.*, 2001, **7**, 1084.

68. P. Lu, T. Teranishi, K. Asakura, M. Miyake and N. Toshima, *J. Phys. Chem. B*, 1999, **103**, 9673.

69. T. Miyake and T. Asakawa, *Appl. Catal. A*, 2005, **280**, 47.

70. T. Teranishi and M. Miyake, *Chem. Mater.*, 1999, **11**, 3414.

71. R. W. J. Scott, H. C. Ye, R. R. Henriquez and R. M. Crooks, *Chem. Mater.*, 2003, **15**, 3873.

72. M. N. Nadagouda and R. S. Varma, *Macromol. Rapid Commun.*, 2007, **28**, 842.

73. Ma, Ming-Guo, *Production of Biofuels and Chemicals with Microwave*, Springer Netherlands, 2015, pp. 169–194.

74. E. D. Sosa, T. K. Darlington, B. A. Hanos and M. J. E. O'Rourke, *J. Compos.*, 2014, 705687, DOI: 10.1155/2014/705687.

75. E. D. Sosa, T. K. Darlington, B. A. Hanos and M. J. E. O'Rourke, *Smart Mater. Res.*, 2015, 248490, DOI: 10.1155/2015/248490.

76. Y. Yang and M. W. Urban, *Chem. Soc. Rev.*, 2013, **42**, 7446.

77. D. Dohler, H. Peterlik and W. H. Binder, *Polymer*, 2015, **69**, 264.

78. Q. L. Tang, K. W. Wang, Y. J. Zhu and F. Chen, *Mater. Lett.*, 2009, **63**, 1332.

79. Y. Kajiwara, A. Nagai and Y. Chujo, *Polym. J.*, 2009, **41**, 1080.

80. P. Cheang and K. A. Khor, *Mater. Sci. Eng. A*, 2003, **345**, 47.

81. A. Guha and S. Nayar, *Bioinspiration Biomimetics*, 2010, **5**, 24001.

82. X. Ma, P. R. Chang, J. Yu and P. Lu, *Carbohydr. Polym.*, 2008, **74**, 895.

83. A. O. Adeodu, I. A. Daniyan, A. H. Oluwole and C. U. Omohimoria, *Int. J. Mater. Sci. Appl.*, 2015, **4**, 229.

84. H. Zeng, C. Gao and D. Yan, *Adv. Funct. Mater.*, 2006, **16**, 812.

85. K. Chrissafis, G. Antoniadis, K. M. Paraskevopoulos, A. Vassiliou and D. N. Bikiaris, *Compos. Sci. Technol.*, 2007, **67**, 2165.

86. X. L. Wang, F. Y. Huang, Y. Zhou and Y. Z. Wang, *J. Macromol. Sci., Part B: Phys.*, 2009, **48**, 710.

87. L. Liao, C. Zhang and S. Gong, *Macromol. Chem. Phys.*, 2007, **208**, 1301.

88. J. Yu, F. Ai, A. Dufresne, S. Gao, J. Huang and P. R. Chang, *Macromol. Mater. Eng.*, 2008, **293**, 763.

89. A. R. Amini, C. T. Laurencin and S. P. Nukavarapu, *Crit. Rev. Biomed. Eng.*, 2012, **40**, 363.

90. M. Vert, M. S. Li, G. Spenlehauer and P. Guerin, *J. Mater. Sci.*, 1992, **3**, 432.

91. I. Zein, D. W. Hutmacher, K. C. Tan and S. H. Teoh, *Biomaterials*, 2002, **23**, 1169.

92. C. X. F. Lam, X. M. Mo, S. H. Teoh and D. W. Hutmacher, *Mater. Sci. Eng., C*, 2002, **20**, 49.

93. P. B. Malafaya, C. Elvira, A. Gallardo, J. S. Roman and R. L. Reis, *J. Biomater. Sci. Polym. Ed.*, 2001, **12**, 1227.

94. V. M. Correlo, M. E. Gomes, K. Tuzlakoglu, J. M. Olivera, P. M. Malafaya, J. F. Mano, N. M. Neves and R. L. Reis, Tissue Engineering Using Natural Polymers, *Biomedical Polymers*, ed. M. Jenkins, CRC Press, Woodhead Publishing Limited, Cambridge, England, 2007, pp. 197–217.

95. F. G. Torres, A. R. Boccaccini and O. P. Troncoso, *J. Appl. Polym. Sci.*, 2007, **103**, 1332.

96. S. Jaya and T. D. Durante, *BioMed. Mater. Eng.*, 2008, **18**, 357.

97. B. T. Ginn and O. Steinbock, *Langmuir*, 2003, **19**, 8117.

98. H. S. Baek, Y. H. Park, C. S. Ki, J. C. Park and D. K. Rah, *Surf. Coat. Techol.*, 2008, **202**, 5794.

99. J. Lee, M. J. Panzer, Y. He, T. P. Lodge and C. D. Frisbie, *J. Am. Chem. Soc.*, 2007, **129**, 4532.

100. J. Lu, F. Yan and J. Texter, *Prog. Polym. Sci.*, 2009, **34**, 431.

101. A. F. Visentin, T. Dong, J. Poli and M. J. Panzer, *J. Mater. Chem. A*, 2014, **2**, 7723.

102. C. Liew, S. Ramesh and A. K. Arof, *Int. J. Hydrogen Energy*, 2014, **39**, 2917.

103. A. I. Horowitz and M. J. Panzer, *J. Mater. Chem.*, 2012, **22**, 16534.

104. K. Ueno, K. Hata, T. Katakabe, M. Kondoh and M. Watanabe, *J. Phys. Chem. B*, 2008, **112**, 9013.

105. A. F. Visentin and M. J. Panzer, *ACS Appl. Mater. Interfaces*, 2012, **4**, 2836.
106. M. A. B. H. Susan, T. Kaneko, A. Noda and M. Watanabe, *J. Am. Chem. Soc.*, 2005, **127**, 4976.
107. X. Zhihong and Z. Yao, ICCM 18, 21–26 August 2011, Korea.
108. F. Zhang, T. Zhou, Y. Liu and J. Leng, *Sci. Rep.*, 2015, 5, 11152, DOI: 10.1038/srep11152.
109. M. D. Hager, S. Bode, C. Weber and U. S. Schubert, *Prog. Polym. Sci.*, 2015, DOI: 10.1016/j.progpolymsci.2015.04.002.
110. P. Spasojevc, B. Adnadevic, S. Velickovic and J. Jovanovic, *J. Appl. Polym. Sci.*, 2011, **119**, 3598.
111. N. Joly, R. Granet, P. Branland, B. Verneuil and P. Krausz, *J. Appl. Polym. Sci*, 2005, **97**, 1266.
112. A. Ebringerova, K. Pielichowski, V. Sasinkova, I. Srokova, V. Tomanova and A. Zoldakova, *Polym. Bull.*, 2008, **60**, 15.
113. I. E. Rivero, V. Balsamo and A. J. Muller, *Carbohydr. Polym.*, 2009, **75**, 343.
114. D. Bogdal, S. Bednarz and K. Matras-Postolek, *Advances in Polymer Science*, Springer, 2015, pp. 1–154.

Subject Index

Page numbers in *italic* refer to figures, schemes or tables.